Algebra through practice

Book 6: Rings, fields and modules

Algebra through practice

A collection of problems in algebra with solutions

Book 6
Rings, fields and modules

T. S. BLYTH ○ E. F. ROBERTSON
University of St Andrews

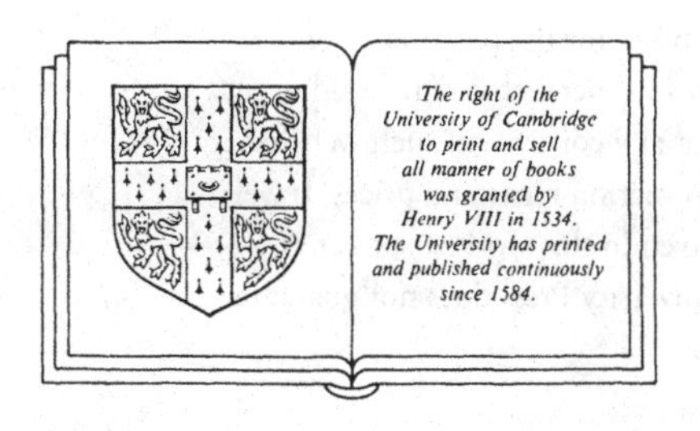

CAMBRIDGE UNIVERSITY PRESS
Cambridge
London New York New Rochelle
Melbourne Sydney

CAMBRIDGE UNIVERSITY PRESS
Cambridge, New York, Melbourne, Madrid, Cape Town,
Singapore, São Paulo, Delhi, Tokyo, Mexico City

Cambridge University Press
The Edinburgh Building, Cambridge CB2 8RU, UK

Published in the United States of America by
Cambridge University Press, New York

www.cambridge.org
Information on this title: www.cambridge.org/9780521272919

First published 1985

A catalogue record for this publication is available from the British Library

Library of Congress Cataloguing in Publication data

ISBN 978-0-521-27291-9 Paperback

Contents

Preface

The aim of this series of problem-solvers is to provide a selection of worked examples in algebra designed to supplement undergraduate algebra courses. We have attempted, mainly with the average student in mind, to produce a varied selection of exercises while incorporating a few of a more challenging nature. Although complete solutions are included, it is intended that these should be consulted by readers only after they have attempted the questions. In this way, it is hoped that the student will gain confidence in his or her approach to the art of problem-solving which, after all, is what mathematics is all about.

The problems, although arranged in chapters, have not been 'graded' within each chapter so that, if readers cannot do problem n this should not discourage them from attempting problem $n+1$. A great many of the ideas involved in these problems have been used in examination papers of one sort or another. Some test papers (without solutions) are included at the end of each book; these contain questions based on the topics covered.

TSB, EFR
St Andrews

Background reference material

Courses on abstract algebra can be very different in style and content. Likewise, textbooks recommended for these courses can vary enormously, not only in notation and exposition but also in their level of sophistication. Here is a list of some major texts that are widely used and to which the reader may refer for background material. The subject matter of these texts covers all six of the present volumes, and in some cases a great deal more. For the convenience of the reader there is given overleaf an indication of which parts of which of these texts are most relevant to the appropriate sections of this volume.

[1] I. T. Adamson, *Introduction to Field Theory*, Cambridge University Press, 1982.
[2] F. Ayres, Jr, *Modern Algebra*, Schaum's Outline Series, McGraw-Hill, 1965.
[3] D. Burton, *A first course in rings and ideals*, Addison-Wesley, 1970.
[4] P. M. Cohn, *Algebra* Vol. I, Wiley, 1982.
[5] D. T. Finkbeiner II, *Introduction to Matrices and Linear Transformations*, Freeman, 1978.
[6] R. Godement, *Algebra*, Kershaw, 1983.
[7] J. A. Green, *Sets and Groups*, Routledge and Kegan Paul, 1965.
[8] I. N. Herstein, *Topics in Algebra*, Wiley, 1977.
[9] K. Hoffman and R. Kunze, *Linear Algebra*, Prentice Hall, 1971.
[10] S. Lang, *Introduction to Linear Algebra*, Addison-Wesley, 1970.
[11] S. Lipschutz, *Linear Algebra*, Schaum's Outline Series, McGraw-Hill, 1974.

[12] I. D. Macdonald, *The Theory of Groups*, Oxford University Press, 1968.
[13] S. MacLane and G. Birkhoff, *Algebra*, Macmillan, 1968.
[14] N. H. McCoy, *Introduction to Modern Algebra*, Allyn and Bacon, 1975.
[15] J. J. Rotman, *The Theory of Groups: An Introduction*, Allyn and Bacon, 1973.
[16] I. Stewart, *Galois Theory*, Chapman and Hall, 1975.
[17] I. Stewart and D. Tall, *The Foundations of Mathematics*, Oxford University Press, 1977.

References useful for Book 6

1: Ideals [**3**, Chapters 2–5], [**4**, Section 10.1], [**8**, Chapter 3], [**13**, Chapter 4].
2: Divisibility [**3**, Chapter 6], [**6**, Chapters 9, 31, 32], [**8**, Chapter 3], [**13**, Chapter 4].
3: Fields [**1**, Chapters 2, 3], [**8**, Chapter 5], [**16**, Chapters 3, 4, 7–12].
4: Modules [**3**, Chapter 12], [**4**, Sections 10.2–10.6], [**13**, Chapter 6].

In [**4**], [**6**] and [**13**] all rings have a 1, and a subring must contain the ring identity. Ring morphisms (called ring homomorphisms in [**4**] and [**6**]) must map the identity element. In [**4**] ring morphisms are written as mappings on the right. The definition of an integral domain given in [**6**] does not require it to be commutative.

1: Ideals

In this chapter we concentrate mainly on ideals of both commutative and non-commutative rings. An ideal I (left, right, or two-sided) of a ring R is maximal if the only ideal (left, right, or two-sided) of R that properly contains I is R itself. An ideal I of R is prime if $I \neq R$ and $xy \in I$ implies $x \in I$ or $y \in I$. These notions are important to the development of ring theory, and in the questions that follow are linked with such ideas as nilpotency and chain conditions. A nilpotent element of a ring R is an element x of R such that $x^n = 0$ for some positive integer n. A nilpotent ring (ideal) is a ring (ideal) every element of which is nilpotent. A ring R satisfies the ascending chain condition on (left, right, two-sided) ideals if every ascending chain

$$I_1 \subseteq I_2 \subseteq I_3 \subseteq \cdots \subseteq I_n \subseteq I_{n+1} \subseteq \cdots$$

of (left, right, two-sided) ideals is such that there exists a positive integer k with $I_n = I_k$ for all $n \geq k$. The descending chain condition is defined similarly.

1.1 Let R be a commutative ring with a 1. If I is an ideal of R prove that

(a) I is prime if and only if R/I is an integral domain;
(b) I is maximal if and only if R/I is a field.

Deduce that every maximal ideal is prime. Give an example of a ring containing an ideal that is prime but not maximal.

Now let R be a non-commutative ring with a 1. Prove that if M is an ideal of R such that every non-zero element of R/M is invertible then M is a maximal ideal. Show that the converse is false by considering the ideal

$$M = \left\{ \begin{bmatrix} 2a & 2b \\ 2c & 2d \end{bmatrix} \mid a, b, c, d \in \mathbb{Z} \right\}$$

in the ring $\mathrm{Mat}_{2\times 2}(\mathbb{Z})$.

1.2 Prove that (8) is a maximal ideal of the ring $4\mathbb{Z}$ but $4\mathbb{Z}/(8)$ is not a field. Explain why this is possible.

1.3 A *boolean ring* is a ring with a 1 in which every element is multiplicatively idempotent (in that $x^2 = x$). Prove that a boolean ring is

(a) of characteristic 2;
(b) commutative.

If I is an ideal of the boolean ring A prove that the following conditions are equivalent :

(1) I is prime;
(2) $A/I \simeq \mathbb{Z}/2\mathbb{Z}$;
(3) I is maximal.

1.4 Let F be a field and let

$$J = \{f \in F[X, Y] \mid f(0, Y) = 0\}.$$

Prove that J is a principal ideal of $F[X, Y]$. Is J prime? Is J maximal?

1.5 Let R be a commutative ring and let A be an ideal of R. Define

$$r(A) = \{x \in R \mid (\exists n \geq 1)\ x^n \in A\}.$$

If I, J are ideals of R prove that

(a) $r(I + J) = r[r(I) + r(J)] \supseteq r(I) + r(J)$;
(b) if $I^n \subseteq J$ for some $n \in \mathbb{N}$ then $r(I) \subseteq r(J)$.

Show that if $r(I) = I$ then R/I has no non-zero nilpotent elements. Is the converse true? Prove that if P is a prime ideal of R then $r(P) = P$.

1.6 If I, J are ideals of a commutative ring R define

$$I : J = \{x \in R \mid xJ \subseteq I\}.$$

Prove that, for ideals I, J, K of R,

(a) if $I \subseteq J$ then $I : K \subseteq J : K$ and $K : I \supseteq K : J$;
(b) $(\forall n \in \mathbb{N})\quad I : J^{n+1} = (I : J^n) : J = (I : J) : J^n$;
(c) $I : J = R \iff J \subseteq I$;
(d) $I : J = I : (I + J)$.

1.7 Let R be a commutative ring with a 1. An ideal Q of R is said to be *primary* if $ab \in Q$ with $a \notin Q$ implies that $b^n \in Q$ for some $n \geq 1$.

Prove that the primary ideals of the ring $\mathbb{Z}$ are precisely the powers of the prime ideals.

Show that $(4, X)$ is a primary ideal of $\mathbb{Z}[X]$ but is not a power of any prime ideal of $\mathbb{Z}[X]$.

Show also that a power of a prime ideal need not be a primary ideal by considering the following example. Let R be the subring of $\mathbb{Z}[X]$ consisting of polynomials in which the coefficient of X is divisible by 3. Show that $P = (3X, X^2, X^3)$ is a prime ideal of R but that P^2 is not primary.

1.8 Let F be a field. Prove that, in $F[X, Y, Z]$,

$$(Y^2Z^2, XYZ) = (Y) \cap (Z) \cap (X, Y)^2 \cap (X, Z)^2.$$

Is it true that $(Y^2Z^2, XYZ) = (Y)(Z)(X, Y)^2(X, Z)^2$? Justify your assertion.

1.9 A set is said to be *inductively ordered* if every non-empty totally ordered subset has an upper bound. *Zorn's axiom* says that every inductively ordered set has a maximal element.

Let A be a ring with a 1 and let $I(A)$ be the set of ideals of A. Given $I \in I(A)$ with $I \neq A$ define

$$F_I = \{J \in I(A) \mid I \subseteq J \subset A\}.$$

Prove that F_I is inductively ordered. By applying Zorn's axiom, deduce that every ideal I of A with $I \neq A$ is contained in a maximal ideal of A. Hence show that if A is commutative then an element of A is a unit if and only if it does not belong to any maximal ideal of A.

1.10 If F is a field prove that the ring $\mathrm{Mat}_{n\times n}(F)$ satisfies both chain conditions on right and left ideals.

Consider the subring

$$R = \left\{ \begin{bmatrix} \alpha & 0 \\ \beta & 0 \end{bmatrix} \mid \alpha, \beta \in \mathbb{R} \right\}$$

of the ring $\mathrm{Mat}_{2\times 2}(\mathbb{R})$. Prove that every proper right ideal of R is of the form

$$\left\{ \alpha \begin{bmatrix} x & 0 \\ y & 0 \end{bmatrix} \mid \alpha \in \mathbb{R} \right\}$$

for fixed $x, y \in \mathbb{R}$. Show that R satisfies both chain conditions on right ideals but neither chain condition on left ideals.

1.11 An element x of a ring R is said to be ***right quasi-regular*** if $x+y+xy=0$ for some $y \in R$.

Prove that $x \in R$ is right quasi-regular if and only if the set $\{r+xr \mid r \in R\}$ coincides with R.

Suppose now that R has a 1. Prove that if x belongs to every maximal right ideal of R then xr is right quasi-regular for every $r \in R$. Show also that if M is a maximal right ideal of R and $x \notin M$ then $-1 = m + xr$ for some $m \in M, r \in R$. Deduce that the intersection of the maximal right ideals of R consists of those $x \in R$ such that, for every $r \in R$, xr is right quasi-regular.

1.12 Let R be a ring and S a subring of R. Suppose that $R^i = S^i + R^{i+1}$ for some $i \in \mathbb{N}$. Prove that $R^j = S^j + R^{j+1}$ for all $j \geq i$. Deduce that, for all $j, k \in \mathbb{N}$ with $j \geq i$, $R^j = S^j + R^{j+k}$.

Suppose now that R is nilpotent. Deduce that

(a) if S is a subring of R with $R^i = S^i + R^{i+1}$ for some $i \in \mathbb{N}$ then $R^j = S^j$ for all $j \geq i$;
(b) if R/R^2 is generated by a single element then so is R;
(c) if M is a maximal ideal of R then $R^2 \subseteq M$;
(d) if M is a maximal ideal of R then R/M has a prime number of elements.

1.13 A ring R is said to be a ***nilring*** if every element of R is nilpotent.

Prove that every subring and every quotient ring of a nilring is also a nilring. Prove also that if A is a two-sided ideal of a ring B with A and B/A both nilrings then B is a nilring. Deduce that the sum of two nil two-sided ideals of a ring is also a nil two-sided ideal.

Let R_p be the set of infinite sequences $a_1, a_2, a_3, \ldots$ where each $a_i \in \mathbb{Z}_{p^i}$ and $a_i = 0$ for all but finitely many suffices i. Given that R_p forms a ring under component-wise addition and multiplication, find all the nilpotent elements of R_p. Prove that this set of nilpotent elements is a nil two-sided ideal that is not nilpotent.

1.14 (a) Let A_1 be an ideal of a ring A_2. Suppose that I is an ideal of A_1 and that J is the smallest ideal of A_2 that contains I. Prove that $J^3 \subseteq I$.

(b) Let R be a ring. Call a subring I of R a ***subideal*** if there is a chain

$$I = A_0 \subseteq A_1 \subseteq A_2 \subseteq \cdots \subseteq A_n = R$$

where A_{i-1} is an ideal of A_i for $i = 1, \ldots, n$. Prove that if I is a subideal of R and $\bar{I}$ is the smallest ideal of R containing I then $\bar{I}^{3n} \subseteq I$.

(c) Deduce from the above that if a subideal I of a ring R is nilpotent then the smallest ideal of R containing I is nilpotent.

1.15 An ideal Q of a ring R is said to be *semiprime* if $A^2 \subseteq Q$ implies $A \subseteq Q$ for every ideal A of R.

Prove that if Q is a semiprime ideal of R and A is an ideal of R such that $A^n \subseteq Q$ for some $n \in \mathbb{N}$ then $A \subseteq Q$.

Let Q be semiprime and let X be a nilpotent ideal of R/Q. If Y is the inverse image of X under the natural morphism $\natural : R \to R/Q$, prove that $Y \subseteq Q$ and deduce that R/Q contains no non-zero nilpotent ideals.

Show conversely that if Q is an ideal of R such that R/Q contains no non-zero nilpotent ideals then Q is semiprime.

2: Divisibility

Here we are mainly concerned with the notion of divisibility in integral domains. If R is an integral domain and $x \in R$ divides $y \in R$ (i.e. there exists $z \in R$ with $y = xz$) then we write $x|y$. If $x|y$ and $y|x$ then x, y are said to be associates. The associates of 1 are called units; they form a group under the multiplication of R. For example, the units of $\mathbb{Z}[\sqrt{n}]$ are the elements of length ± 1 where the length of $a + b\sqrt{n}$ is $\ell(a + b\sqrt{n}) = a^2 - nb^2$.

The usual notion of a prime integer gives rise in an arbitrary integral domain to the two distinct concepts of irreducible element and prime element. An element x is irreducible if it is neither zero nor a unit and its only divisors are its associates or units. We say that x is prime if it is neither zero nor a unit and is such that if it divides a product then it divides one of the factors. These two concepts coincide in the case of a principal ideal domain (i.e. an integral domain in which every ideal is principal).

Associated with the idea of divisibility is that of highest common factor (h.c.f.) or greatest common divisor (g.c.d.), and least common multiple (l.c.m.). These need not exist in general, but are important in euclidean domains (i.e. an integral domain D with a function $N : D \setminus \{0\} \to \mathbb{N}$, called a norm function, such that if $b|a$ then $N(b) \leq N(a)$ and if $a, b \neq 0$ then $a = bq + r$ where $r = 0$ or $N(r) < N(b)$).

2.1 Let x, y, u be elements of an integral domain R. Prove that

(a) $x|y$ if and only if $(y) \subseteq (x)$;
(b) x, y are associates if and only if $(x) = (y)$;
(c) u is a unit if and only if $(u) = R$;
(d) y is a proper factor of x if and only if $(x) \subset (y) \subset R$;

(e) x is irreducible if and only if (x) is maximal among the principal ideals of R.

2.2 In the following statements R is an integral domain and $R^\star = R \setminus \{0\}$. Give a proof for each statement that is true and a counter-example for each that is false.

(a) If $|$ is an equivalence relation on $R^\star$ then R is a field.
(b) If $x \sim y$ for all $x, y \in R^\star$ then R is a field.
(c) If $a \sim b$ and $c \sim d$ then $ac \sim bd$.
(d) If $a \sim b$ and $c \sim d$ then $(a+c) \sim (b+d)$.
(e) If every element of $R^\star$ is a unit or a prime then R is a field.

2.3 Determine whether 5 is irreducible in each of

$$\mathbb{Z},\quad \mathbb{Z}[X],\quad \mathbb{Z}[i],\quad \mathbb{Z}[\sqrt{-2}].$$

2.4 Write $43i - 19$ as a product of irreducibles in $\mathbb{Z}[i]$.

2.5 Express as products of irreducibles

(a) $11 + 7\sqrt{-1}$ in $\mathbb{Z}[\sqrt{-1}]$;
(b) $4 + 7\sqrt{2}$ in $\mathbb{Z}[\sqrt{2}]$;
(c) $4 - \sqrt{-3}$ in $\mathbb{Z}[\sqrt{-3}]$.

2.6 (a) Does $\sqrt{6} \cdot \sqrt{6} = 3 \cdot 2$ violate unique factorisation in $\mathbb{Z}[\sqrt{6}]$?
(b) Show that $\mathbb{Z}[\sqrt{10}]$ is not a unique factorisation domain.

2.7 Prove that $\mathbb{Z}[\sqrt{-6}]$ is not a unique factorisation domain by finding two different factorisations of 10.

Find examples of each of the following in $\mathbb{Z}[\sqrt{-6}]$ and justify your assertions:

(a) an irreducible element that is not prime;
(b) non-zero elements a, b such that g.c.d.(a, b) does not exist;
(c) non-zero elements a, b for which g.c.d.$(a, b) = 1$ but no α, β exist in $\mathbb{Z}[\sqrt{-6}]$ such that $\alpha a + \beta b = 1$.

2.8 Prove that in $\mathbb{Z}[\sqrt{-7}]$ the element 8 can be written both as a product of two irreducible elements and as a product of three irreducible elements. Deduce that for every positive integer k there is an element of $\mathbb{Z}[\sqrt{-7}]$ that can be written as products of $t+1, t+2, \ldots, t+k$ irreducible elements for some integer t.

Give an example of an irreducible element of $\mathbb{Z}[\sqrt{-7}]$ that is not prime. Give an example of elements $\alpha, \beta \in \mathbb{Z}[\sqrt{-7}]$ such that g.c.d.$(\alpha, \beta) = 1$ but no $\gamma, \delta \in \mathbb{Z}[\sqrt{-7}]$ exist with $\gamma\alpha + \delta\beta = 1$. Justify your assertions.

2.9 If I is a non-zero ideal of $\mathbb{Z}[i]$, use the fact that I is principal together with the euclidean algorithm to show that $\mathbb{Z}[i]/I$ has only finitely many elements.

2.10 Let $\alpha \in \mathbb{Z}[i]$ be prime. Prove that precisely one of the following statements holds :

(a) $\alpha \sim 1+i$;
(b) $\alpha \in \mathbb{Z}$ is prime and $\alpha \equiv 3 \bmod 4$;
(c) α divides a prime $p \in \mathbb{Z}$ with $p \equiv 1 \bmod 4$.

2.11 For the quadratic domain $\mathbb{Z}[\sqrt{n}]$ where n is square-free prove that

(a) if $n < -1$ then the group of units of $\mathbb{Z}[\sqrt{n}]$ is $\{-1, 1\}$;
(b) if $n > 1$ and the group of units of $\mathbb{Z}[\sqrt{n}]$ contains more than two elements then it is infinite;
(c) the group of units of $\mathbb{Z}[\sqrt{2}]$ is $\{\pm(1+\sqrt{2})^k \mid k \geq 1\}$.

2.12 Let $p \in \mathbb{Z}$ be prime and consider the subring of $\mathbb{Q}$ given by

$$R = \{\frac{m}{n} \mid m, n \in \mathbb{Z},\ n \neq 0,\ p \text{ does not divide } n\}.$$

Prove that every ideal of R is principal and that R contains a unique maximal ideal.

2.13 Let D be a principal ideal domain. Prove that

(a) if $x, y_1, \ldots, y_n \in D \backslash \{0\}$ are such that x and y_i are relatively prime for each i then x and $y_1 \cdots y_n$ are also relatively prime;

(b) if $p_1, \ldots, p_n$ are irreducible elements of D, no pair of which are associates, then for all positive integers $r_1, \ldots, r_n$ the elements $p_1^{r_1}, \ldots, p_n^{r_n}$ are pairwise relatively prime.

Hence show that if F is the field of quotients of D and $x = a/b \in F$ where $b = p_1^{r_1} \cdots p_n^{r_n}$ is the unique factorisation of b as a product of irreducibles, no two distinct p_i being associates, then there exist $a_1, \ldots, a_n \in D$ such that

$$x = \frac{a_1}{p_1^{r_1}} + \cdots + \frac{a_n}{p_n^{r_n}}.$$

2.14 If A is a commutative ring with a 1 then A is said to be

(a) *noetherian* if every ascending chain of ideals of A terminates finitely;

(b) *satisfy the maximum condition* if every family of ideals of A has a maximal element.

Prove that the following are equivalent :

(1) A is noetherian;
(2) A satisfies the maximum condition;
(3) every ideal of A is finitely generated.

Deduce that if A is an integral domain then the following are equivalent :

(α) A is a principal ideal domain;

(β) A is noetherian and the sum of any two principal ideals is a principal ideal.

2.15 If R is a euclidean domain with norm N, prove that

(a) if $a|b$ and $N(a) = N(b)$ then a and b are associates;

(b) if a, b are non-zero elements of R neither of which divides the other then there exist $\alpha, \beta, d \in R$ with $\alpha a + \beta b = d$ and $N(d) < \min(N(a), N(b))$.

2.16 Prove that in the definition of a euclidean norm δ quotients and remainders are unique if and only if

$$\delta(a + b) \leq \max(\delta(a), \delta(b)).$$

Give an example in $\mathbb{Z}[i]$ where quotients and remainders are not unique.

2.17 (a) Write $-1 + 3i$ as a product of primes in $\mathbb{Z}[i]$.

(b) Let R be a unique factorisation domain and let

$$f(X) = a_0 + a_1 X + a_2 X^2 + \cdots + a_n X^n \in R[X]$$

be such that g.c.d.$(a_0, \ldots, a_n) = 1$. Suppose that there is a prime $p \in R$ such that $p|a_i$ for $i = 0, \ldots, n-1$, p^2 does not divide a_0, and p does not divide a_n. Prove that $f(X)$ is irreducible in $R[X]$.

(c) Deduce from (a) and (b) that

$$X^3 + 8iX^2 - 6X - 1 + 3i$$

is irreducible in $\mathbb{Z}[i][X]$.

2.18 Does every pair of non-zero elements of $\mathbb{Z}[\sqrt{3}]$ have a greatest common divisor? Find the greatest common divisor of 13 and $7 + 5\sqrt{3}$.

2.19 In the ring $\mathbb{Z}[\sqrt{-5}]$ prove that

(a) the units are 1 and -1;

(b) $3, 2 + \sqrt{-5}, 2 - \sqrt{-5}$ are irreducible;

(c) 9 has two different factorisations into a product of irreducibles;

(d) the ideals $(3, 2 + \sqrt{-5})$ and $(3, 2 - \sqrt{-5})$ are prime;

(e) although 3 is irreducible the ideal (3) can be expressed as a product of prime ideals.

2.20 Consider the following ideals in $\mathbb{Z}[\sqrt{10}]$:

$$P_1 = (2, \sqrt{10}), \quad P_2 = (3, 4 + \sqrt{10}), \quad P_3 = (3, 4 - \sqrt{10}).$$

Show that each of these ideals is prime. Prove that the two prime decompositions

$$6 = 2 \cdot 3 = (4 + \sqrt{10})(4 - \sqrt{10})$$

lead to the same decomposition of (6) as products of the prime ideals P_1, P_2, P_3.

3: Fields

This chapter is mainly concerned with the notion of a field extension; i.e. an embedding of one field into another. If K is an extension of F then K can be considered as a vector space over F. The dimension of this vector space is called the degree of the extension and is written $(K : F)$. When this is finite, the extension is said to be finite. An element a of K is said to be algebraic over F if $p(a) = 0$ for some $p(X) \in F[X]$. The minimum polynomial of a is the monic polynomial of least degree with this property. An extension K of F is said to be algebraic if every element of K is algebraic over F; otherwise K is said to be transcendental.

When K is an extension of F and $a_1, \ldots, a_n \in K$ we use the (standard) notation $F(a_1, \ldots, a_n)$ to denote the smallest subfield of K that contains $F \cup \{a_1, \ldots, a_n\}$. In particular, $F(a)$ is called a simple extension of F. A field K is called a splitting field of $f(X) \in F[X]$ if K is an extension of F of minimal degree in which f can be expressed as a product of linear factors. K is called a normal extension of F if it is the splitting field of some $f(X) \in F[X]$.

Finally, we assume that the reader is familiar with the statement of the fundamental theorem of Galois theory, which relates subgroups of the Galois group $\mathrm{Gal}(K, F)$ with subfields of K containing F.

3.1 Let F be a field of characteristic p. Prove that

$$(\forall a, b \in F) \qquad (a \pm b)^p = a^p \pm b^p.$$

Hence deduce by induction that, for every positive integer n,

$$(\forall a, b \in F) \qquad (a \pm b)^{p^n} = a^{p^n} \pm b^{p^n}.$$

3.2 Let F be a field and $D : F[X] \to F[X]$ the differentiation map described by

$$D(a_0 + a_1X + a_2X^2 + \cdots + a_nX^n) = a_1 + 2a_2X + \cdots + na_nX^{n-1}.$$

If F is of characteristic 0 prove that $Df = 0$ if and only if f is a constant polynomial. If F is of characteristic p prove that $Df = 0$ if and only if f is of the form

$$a_0 + a_pX^p + \cdots + a_{rp}X^{rp}.$$

3.3 If F is a field prove that the units of $F[X]$ are the non-zero constant polynomials.

Let $i : F \to F[X]$ be the canonical injection. Prove that if $\varphi : F[X] \to F[X]$ is an automorphism then there is an automorphism $\vartheta : F \to F$ such that the diagram

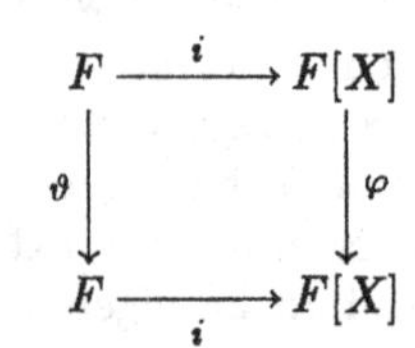

is commutative (in the sense that $\varphi \circ i = i \circ \vartheta$). Deduce that there exist $a, b \in F$ with $a \neq 0$ such that $\varphi(X) = aX + b$.

3.4 Let F be a field. Given $f, g \in F[X]$ we say that f, g are *equivalent* if $f(\alpha) = g(\alpha)$ for every $\alpha \in F$. If F is infinite, prove that f, g are equivalent if and only if $f = g$. If F is finite, say $F = \{a_1, \ldots, a_n\}$, prove that f, g are equivalent if and only if $f - g$ is divisible by

$$m(X) = (X - a_1)(X - a_2)\cdots(X - a_n).$$

In the case where $F = \mathrm{GF}(p) = \mathbb{Z}/p\mathbb{Z}$, show that $m(X) = X^p - X$.

3.5 If the field F is not of characteristic 2, prove that $X^2 + Y^2 - 1$ is irreducible in $F[X, Y]$.

3.6 By considering the ring morphism $\mathbb{Z}[X] \to \mathbb{Z}_n[X]$ that is induced by the natural morphism $\mathbb{Z} \to \mathbb{Z}_n$, show that if f is irreducible over $\mathbb{Z}_n$ then it is irreducible over $\mathbb{Z}$.

By taking $n = 5$, show that if

$$f(X) = X^4 + 15X^3 + 7$$

then f is irreducible over $\mathbb{Q}$.

3.7 Determine a basis for $\mathbb{Q}(\sqrt{2}, \sqrt[3]{2})$ over $\mathbb{Q}$. Deduce that $\sqrt[6]{2} \in \mathbb{Q}(\sqrt{2}, \sqrt[3]{2})$. Hence show that $\mathbb{Q}(\sqrt{2}, \sqrt[3]{2})$ is a simple extension of $\mathbb{Q}$.

3.8 Let F be a field and consider

$$Z = \frac{X^3}{X+1} \in F(X).$$

Prove that Z is transcendental over F but that $F(X)$ is a simple algebraic extension of $F(Z)$. What is the minimum polynomial of X over $F(Z)$?

3.9 Show that $X^2 - 3$ and $X^2 - 2X - 2$ are both irreducible over $\mathbb{Q}$ and have the same splitting field.

3.10 Show that $(X^2 - 2X - 2)(X^2 + 1)$ and $X^5 - 3X^3 + X^2 - 3$ have the same splitting field K over $\mathbb{Q}$ and find $(K : \mathbb{Q})$.

3.11 Let K be a field and let $a, b \in K$. Prove that $X + a + b$ divides $X^3 - 3abX + a^3 + b^3$ in the ring $K[X]$, and determine $q(X) \in K[X]$ such that

$$X^3 - 3abX + a^3 + b^3 = (X + a + b)q(X).$$

Find a splitting field S for $X^6 - 6X^3 + 8$ over $\mathbb{Q}$. What is the degree of S over $\mathbb{Q}$? Show that S contains the element $\sqrt[3]{4} + \sqrt[3]{2}$. Find the minimum polynomial of this element over $\mathbb{Q}$.

Is it true that $\mathbb{Q}(\sqrt[3]{4} + \sqrt[3]{2})$ is a splitting field for $X^6 - 6X^3 + 8$?

3.12 Find quadratic factors for

$$f(X) = X^4 + 2X^3 - 8X^2 - 6X - 1$$

in $\mathbb{Q}[X]$ and hence show that $\mathbb{Q}(\sqrt{2}, \sqrt{3})$ is a splitting field for f over $\mathbb{Q}$.

Show that $\sqrt{3} \notin \mathbb{Q}(\sqrt{2})$. Deduce that $(\mathbb{Q}(\sqrt{2}, \sqrt{3}) : \mathbb{Q}) = 4$ and find a basis for $\mathbb{Q}(\sqrt{2}, \sqrt{3})$ over $\mathbb{Q}$.

Prove that $\mathbb{Q}(\sqrt{2}, \sqrt{3}) = \mathbb{Q}(\sqrt{3} - \sqrt{2})$ and find the minimum polynomial of $\sqrt{3} - \sqrt{2}$ over $\mathbb{Q}(\sqrt{3})$.

3.13 Find the irreducible factors in $\mathbb{Q}[X]$ of

$$f(X) = X^4 - X^2 - 2.$$

Show that $\mathbb{Q}(i, \sqrt{2})$ is a splitting field for f over $\mathbb{Q}$. Show also that $(\mathbb{Q}(i, \sqrt{2}) : \mathbb{Q}) = 4$ and write down a basis for $\mathbb{Q}(i, \sqrt{2})$ over $\mathbb{Q}$. Find the minimum polynomial of $i+\sqrt{2}$ over $\mathbb{Q}$. Deduce that $(\mathbb{Q}(i+\sqrt{2}) : \mathbb{Q}) = 4$ and hence that $\mathbb{Q}(i, \sqrt{2}) = \mathbb{Q}(i + \sqrt{2})$.

Noting that $\mathbb{Q}(i), \mathbb{Q}(\sqrt{2})$ and $\mathbb{Q}(i\sqrt{2})$ are all subfields of $\mathbb{Q}(i+\sqrt{2})$ over each of which the degree of $\mathbb{Q}(i+\sqrt{2})$ is 2, find the minimum polynomial of $i + \sqrt{2}$ over each of

(a) $\mathbb{Q}(i)$;
(b) $\mathbb{Q}(\sqrt{2})$;
(c) $\mathbb{Q}(i\sqrt{2})$.

3.14 Given $r \in \mathbb{Q}$ consider the polynomial $f(X) = X^4 + r \in \mathbb{Q}[X]$. Prove that the following statements are equivalent :

(1) f is reducible;
(2) f has a quadratic factor;
(3) either $r = -p^2$ $(p \in \mathbb{Q})$ or $r = \frac{1}{4}q^4$ $(q \in \mathbb{Q})$.

Suppose now that f is irreducible over $\mathbb{Q}$. Let $\xi \in \mathbb{C}$ be a root of f and let $K = \mathbb{Q}(\xi)$. Prove that

(a) if $\sqrt{r} \notin \mathbb{Q}$ then the only subfield of K, other than K and $\mathbb{Q}$, is $\mathbb{Q}(\xi^2)$;

(b) if $\sqrt{r} \in \mathbb{Q}$ then the subfields of K, other than K and $\mathbb{Q}$, are

$$\mathbb{Q}(\xi^2), \quad \mathbb{Q}(\sqrt{r}\xi + \xi^3), \quad \mathbb{Q}(-\sqrt{r}\xi + \xi^3).$$

3.15 Determine the order of the Galois group of $\mathbb{Q}(\omega)$ over $\mathbb{Q}$ when

(a) $\omega = e^{2\pi i/5}$;
(b) $\omega = \sqrt[3]{2}$.

3.16 Find a splitting field for the polynomial $X^4 - 2$ over $\mathbb{Q}$. Show that the Galois group of $X^4 - 2$ is non-abelian and of order 8.

3.17 Determine a subfield of $\mathbb{C}$ that is a splitting field over $\mathbb{Q}$ for the polynomial $X^3 - 2$. Construct the Galois group of this polynomial and identify the subfields of the splitting field.

3.18 Show that if α is a root of

$$X^3 - 3X + 1 = 0$$

then $\alpha^2 - 2$ and $2 - \alpha - \alpha^2$ are the other roots.

Let $f(X) = X^3 - 3X + 1$. Show that $\mathbb{Q}(\alpha)$ is a splitting field for f over $\mathbb{Q}$ and find $(\mathbb{Q}(\alpha) : \mathbb{Q})$.

Show that there is a unique $\mathbb{Q}$–automorphism ϑ of $\mathbb{Q}(\alpha)$ with $\vartheta(\alpha) = \alpha^2 - 2$ and determine $\mathrm{Gal}(\mathbb{Q}(\alpha), \mathbb{Q})$. Is it true that every permutation of the roots of f extends to an element of $\mathrm{Gal}(\mathbb{Q}(\alpha), \mathbb{Q})$?

3.19 Determine which of the following extensions of $\mathbb{Q}$ are normal :

(a) $\mathbb{Q}(\sqrt{2})$;
(b) $\mathbb{Q}(\sqrt[3]{2})$;
(c) $\mathbb{Q}(\sqrt[4]{2})$;
(d) $\mathbb{Q}(\sqrt[5]{2})$.

3.20 Let E be a field of characteristic 0 and let $K = E(X_1, \ldots, X_n)$ be the field of rational functions in n indeterminates over E. Define the *elementary symmetric polynomials* in $X_1, \ldots, X_n$ by

$$\sigma_1 = \sum_{i=1}^{n} X_i, \quad \sigma_2 = \sum_{i \neq j} X_i X_j, \quad \ldots, \quad \sigma_n = X_1 X_2 \cdots X_n.$$

Let $F = E(\sigma_1, \ldots, \sigma_n)$ and let

$$f(X) = X^n - \sigma_1 X^{n-1} + \sigma_2 X^{n-2} - \cdots + (-1)^n \sigma_n \in F[X].$$

Prove that

(a) K is a splitting field for f over F;
(b) the Galois group of K over F is isomorphic to the symmetric group of degree n;
(c) $(K : F) = n!$.

4: Modules

The easiest way to define a module is to say that it is an algebraic system that satisfies the same axioms as a vector space except that the scalars come from a ring R with a 1 instead of from a field F. This seemingly modest generalisation leads to an algebraic structure that is of the greatest importance. We use here the term R–module, it being understood that the scalars are written on the left.

As with groups and rings, the substructures (submodules) and the structure-preserving mappings (R–morphisms) are of especial interest, as are 'new' structures obtained from 'old' such as quotient modules, cartesian (or direct) products, and direct sums. We assume that the reader is familiar with the correspondence theorem and the basic isomorphism theorems. If M, N are R–modules then the set of R–morphisms $f : M \to N$ forms an additive abelian group (a $\mathbb{Z}$–module) which we denote by $\mathrm{Mor}_R(M, N)$, or $\mathrm{End}_R(M)$ in the case where $N = M$. Some authors write the former as $\mathrm{Hom}_R(M, N)$.

Topics dealt with include exact sequences (the image of the input R–morphism equals the kernel of the output R–morphism), various diagrams that are commutative (all composite R–morphisms defined from one given R–module to another are equal), the chain conditions (as for rings in Chapter 1), and Jordan–Hölder towers of submodules (as for composition series in groups). We also use the notion of a free R–module (one that has a basis), and a projective R–module (a direct summand of a free R–module).

4.1 Let M be an abelian group and let $\mathrm{End}\, M$ be the ring of endomorphisms on M (i.e. group morphisms $f : M \to M$ under addition and composition). Prove that M is an $(\mathrm{End}\, M)$–module under the external law

$\operatorname{End} M \times M \to M$ described by $(f, m) \mapsto fm = f(m)$.

4.2 Let R be a ring with a 1, and M an abelian group. Prove that M is an R–module if and only if there is a 1–preserving ring morphism $\mu : R \to \operatorname{End} M$.

4.3 Prove that the ring of endomorphisms of the abelian group $\mathbb{Z}$ is isomorphic to the ring $\mathbb{Z}$; and that the ring of endomorphisms of the abelian group $\mathbb{Q}$ is isomorphic to the field $\mathbb{Q}$.

4.4 Let R be a commutative ring with a 1 and let $f : R \times R \to R$ be a mapping. Prove that f is an R–morphism if and only if there exist $\alpha, \beta \in R$ such that

$$(\forall x, y \in R) \qquad f(x, y) = \alpha x + \beta y.$$

4.5 Let $\mathbb{Z}[\sqrt{2}] = \{a+b\sqrt{2} \mid a, b \in \mathbb{Z}\}$ and consider the mapping $f : \mathbb{Z}[\sqrt{2}] \to \mathbb{Z}[\sqrt{2}]$ given by

$$f(a + b\sqrt{2}) = a + b.$$

Determine whether or not f is

(a) a ring morphism;
(b) a $\mathbb{Z}[\sqrt{2}]$–morphism;
(c) a $\mathbb{Z}$–morphism.

4.6 Let M be an R–module. For every R–morphism $f : R \to M$ and every $\lambda \in R$ let $\lambda f : R \to M$ be given by $(\lambda f)(r) = f(r\lambda)$. Prove that $\lambda f \in \operatorname{Mor}_R(R, M)$ and deduce that $\operatorname{Mor}_R(R, M)$ is an R–module. Show also that the mapping $\vartheta : \operatorname{Mor}_R(R, M) \to M$ given by $\vartheta(f) = f(1_R)$ is an R–isomorphism.

4.7 Let $f : M \to N$ be an R–morphism. If A is a submodule of M let $f^{\rightarrow}(A) = \{f(a) \mid a \in A\}$, and if B is a submodule of N let $f^{\leftarrow}(B) = \{x \in M \mid f(x) \in B\}$. Prove that

(a) $f^{\leftarrow}[f^{\rightarrow}(A)] = A + \operatorname{Ker} f$;
(b) $f^{\rightarrow}[f^{\leftarrow}(B)] = B \cap \operatorname{Im} f$;
(c) $f^{\rightarrow}(A \cap f^{\leftarrow}(B)) = f^{\rightarrow}(A) \cap B$.

4.8 An R–monomorphism $f : M \to N$ is said to be *essential* if, for every non-zero submodule A of N, $f^{\leftarrow}(A)$ is a non-zero submodule of M. If M is a submodule of N then N is said to be an *essential extension* of M if the canonical inclusion $i : M \to N$ is essential.

Prove that, as $\mathbb{Z}$–modules,

(a) $\mathbb{Q}$ is an essential extension of $\mathbb{Z}$;
(b) $\mathbb{R}$ is not an essential extension of $\mathbb{Q}$.

If M is a submodule of N, use Zorn's axiom to prove that there is a submodule A of N that is maximal with respect to the property $M \cap A = 0$. For such a submodule A prove that the composite morphism

$$M \xrightarrow{i} N \xrightarrow{\natural} N/A$$

is an essential monomorphism.

4.9 Given the sequence

$$X \xrightarrow{f} Y \xrightarrow{g} Z$$

of R-modules and R-morphisms, prove that

(a) $\operatorname{Im} f \cap \operatorname{Ker} g = f^{\rightarrow}(\operatorname{Ker} g \circ f)$;
(b) $\operatorname{Im} f + \operatorname{Ker} g = g^{\leftarrow}(\operatorname{Im} g \circ f)$.

4.10 Given the diagram

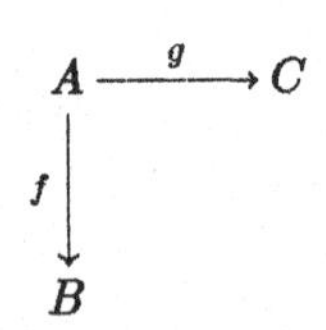

of R-modules and R-morphisms in which f is an epimorphism, prove that the following statements are equivalent :

(a) there is an R-morphism $h : B \to C$ such that $h \circ f = g$;
(b) $\operatorname{Ker} f \subseteq \operatorname{Ker} g$.

Prove further that such an R-morphism h, when it exists, is unique; and that it is a monomorphism if and only if $\operatorname{Ker} f = \operatorname{Ker} g$.

If an R-morphism $\vartheta : M \to N$ can be expressed as a composite morphism

$$M \xrightarrow{\alpha} A \xrightarrow{\beta} B \xrightarrow{\gamma} N$$

where α is an epimorphism, β is an isomorphism, and γ is a monomorphism, prove that $A \simeq M/\operatorname{Ker} \vartheta$ and $B \simeq \operatorname{Im} \vartheta$.

4.11 Establish the following statements concerning the various $\mathbb{Z}$-modules involved :

(a) $\mathbb{Q}$ is not finitely generated;
(b) $\mathbb{Q}/\mathbb{Z}$ is infinite;
(c) $\emptyset$ is the only independent subset of $\mathbb{Q}/\mathbb{Z}$;
(d) $\operatorname{Mor}_{\mathbb{Z}}(\mathbb{Z}/2\mathbb{Z}, \mathbb{Q}) = 0$;
(e) $\operatorname{Mor}_{\mathbb{Z}}(\mathbb{Q}, \mathbb{Z}) = 0$.

4.12 An R–module M is said to be *cyclic* if it is generated by a singleton subset. Let $M = Rx$ be a cyclic R–module and let $\text{Ann}_R(x) = \{\lambda \in R \mid \lambda x = 0\}$. Prove that $\text{Ann}_R(x)$ is a submodule of R and that $M \simeq R/\text{Ann}_R(x)$.

Let m and n be integers, each greater than 1. Show that the prescription $\vartheta(x+m\mathbb{Z}) = nx+nm\mathbb{Z}$ describes a $\mathbb{Z}$–morphism $\vartheta : \mathbb{Z}/m\mathbb{Z} \to \mathbb{Z}/nm\mathbb{Z}$. Prove that the $\mathbb{Z}$–module $\text{Mor}_{\mathbb{Z}}(\mathbb{Z}/m\mathbb{Z}, \mathbb{Z}/nm\mathbb{Z})$ is generated by $\{\vartheta\}$ and deduce that

$$\text{Mor}_{\mathbb{Z}}(\mathbb{Z}/m\mathbb{Z}, \mathbb{Z}/nm\mathbb{Z}) \simeq \mathbb{Z}/m\mathbb{Z}.$$

4.13 If A, B are submodules of an R–module M, establish a short exact sequence

$$0 \longrightarrow A \cap B \longrightarrow A \times B \longrightarrow A + B \longrightarrow 0.$$

4.14 The diagram of R–modules and R–morphisms

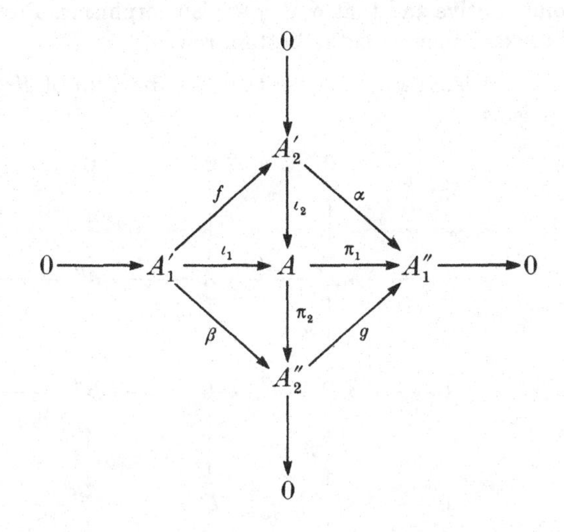

is given to be commutative with the row and column exact. Prove that

(a) α and β are zero morphisms;

(b) f and g are isomorphisms.

4.15 Let M and N be R–modules and $f : M \to N$ an R–morphism. If A is a submodule of M prove that the following statements are equivalent :

(a) there is a unique R–morphism $f_\star : M/A \to N$ such that $f_\star \circ \natural_A = f$;

(b) $A \subseteq \text{Ker}\, f$.

Show further that such an R–morphism $f_\star$ is a monomorphism if and only if $A = \operatorname{Ker} f$.

If A, B are submodules of an R-module M, establish an exact sequence of the form

$$0 \longrightarrow M/(A \cap B) \longrightarrow M/A \times M/B \longrightarrow M/(A+B) \longrightarrow 0.$$

Hence deduce that

$$(A+B)/(A\cap B) \simeq (A+B)/A \times (A+B)/B.$$

4.16 Suppose that the diagram of R–modules and R–morphisms

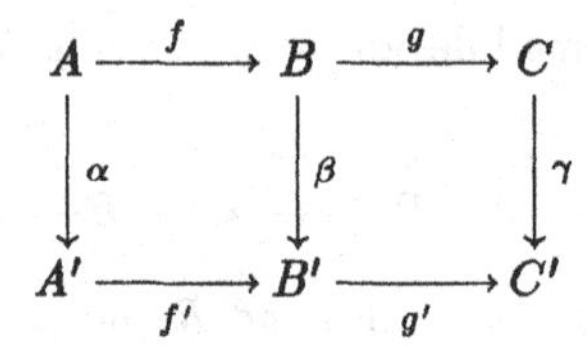

is commutative and that α, β, γ are isomorphisms. Prove that if the top row is exact then so is the bottom row.

4.17 [The 3×3 lemma.] Suppose that the diagram of R–modules and R–morphisms

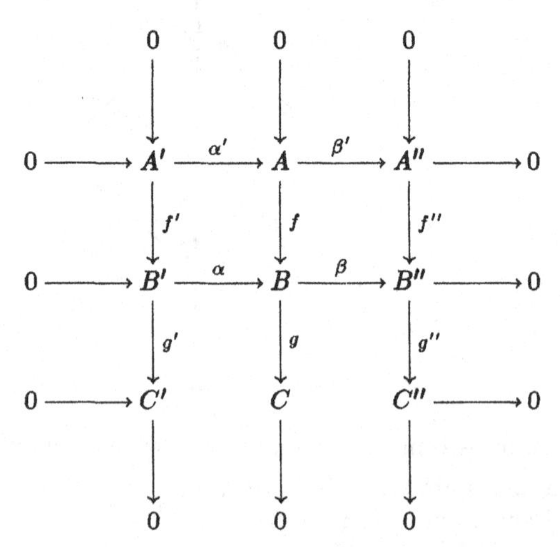

is commutative, that all three columns are exact, and that the top two rows are exact. Prove that there exist unique R–morphisms $\alpha'' : C' \to C$ and $\beta'' : C \to C''$ such that the resulting bottom row is exact and the completed diagram is commutative.

4.18 Let R be an integral domain. Given $x \in R$ with $x \neq 0$, show that there is a descending chain

$$R \supseteq Rx \supseteq Rx^2 \supseteq \cdots \supseteq Rx^n \supseteq Rx^{n+1} \supseteq \ldots$$

of submodules of the R-module R such that, for every n,

$$Rx^n/Rx^{n+1} \simeq R/Rx.$$

4.19 Determine which of the chain conditions, if any, are satisfied in each of the following modules.

(a) $\mathbb{Z}$ as a $\mathbb{Z}$-module;
(b) $\mathbb{Z}_m$ as a $\mathbb{Z}$-module;
(c) $\mathbb{Z}_m$ as a $\mathbb{Z}_m$-module;
(d) $\mathbb{Q}$ as a $\mathbb{Q}$-module;
(e) $\mathbb{Q}$ as a $\mathbb{Z}$-module;
(f) $\mathbb{Q}[X]$ as a $\mathbb{Q}$-module;
(g) $\mathbb{Q}[X]$ as a $\mathbb{Q}[X]$-module;
(h) $\mathbb{Q}[X]/M$ as a $\mathbb{Q}[X]$-module where M is the submodule consisting of those polynomials divisible by X^5;
(i) $\mathbb{Q}[X]/M$ as a $\mathbb{Q}[X]$-module where M is the submodule consisting of those polynomials divisible by $X^2 + 1$.

4.20 Let M be an R-module of finite height. If N is a submodule of M prove that there is a Jordan–Hölder tower of submodules

$$M = M_0 \supset M_1 \supset \cdots \supset M_{r-1} \supset M_r = 0$$

such that, for some index k, $M_k = N$.

4.21 If M is an R-module of finite height and if N is a submodule of M prove that $h(M) = h(N) + h(M/N)$. Deduce that $N = M$ if and only if $h(N) = h(M)$.

4.22 If M and N are R-modules with M of finite height and if $f : M \to N$ is an R-morphism, prove that $\operatorname{Im} f$ and $\operatorname{Ker} f$ are of finite height with

$$h(\operatorname{Im} f) + h(\operatorname{Ker} f) = h(M).$$

4.23 Let $M_1, \ldots, M_n$ be R-modules each of finite height. If

$$0 \longrightarrow M_1 \xrightarrow{f_1} M_2 \xrightarrow{f_2} \ldots \xrightarrow{f_{n-1}} M_n \longrightarrow 0$$

is an exact sequence prove that

$$\sum_{k=1}^{n} (-1)^k h(M_k) = 0.$$

4.24 Find $\mathbb{Z}$-modules M, N with $h(M) = h(N) = 2$ and $\operatorname{Mor}_{\mathbb{Z}}(M, N) = 0$.

4.25 Let F be a field and let M_n be the ring of $n \times n$ matrices over F. For $i = 1, \ldots, n$ let $E_i \in M_n$ be the matrix whose (i,i)-th entry is 1 and all other entries are 0. For $i = 1, \ldots, n$ define

$$B_i = M_n(E_1 + \cdots + E_i).$$

Prove that

$$M_n = B_n \supset B_{n-1} \supset \cdots \supset B_1 \supset B_0 = 0$$

is a Jordan–Hölder tower for the M_n-module M_n.

4.26 Let $M_1, \ldots, M_n$ be submodules of an R-module M such that $M = \bigoplus_{i=1}^n M_i$. For $k = 1, \ldots, n$ let N_k be a submodule of M_k. If $N = \sum_{i=1}^n N_i$ prove that $N = \bigoplus_{i=1}^n N_i$ and that $M/N = \bigoplus_{i=1}^n M_i/N_i$.

4.27 If $A_1, \ldots, A_n$ are submodules of the R-module M prove that $M = \bigoplus_{i=1}^n A_i$ if and only if for $i = 1, \ldots, n$ there exist R-morphisms $f_i : A_i \to M$ and $g_i : M \to A_i$ such that

(1) $g_i \circ f_j = \begin{cases} \mathrm{id}_{A_i} & \text{if } i = j, \\ 0 & \text{if } i \neq j; \end{cases}$

(2) $\sum_{i=1}^n f_i \circ g_i = \mathrm{id}_M$.

Deduce that M is the direct sum of submodules M_1, M_2 if and only if there is a split short exact sequence

$$0 \longrightarrow M_1 \longrightarrow M \longrightarrow M_2 \longrightarrow 0.$$

4.28 Let A be an R-module. Use the fact that R has an identity element to prove that every exact sequence of R-modules and R-morphisms of the form

$$0 \longrightarrow X \longrightarrow A \longrightarrow R \longrightarrow 0$$

splits. In contrast, exhibit an exact sequence of $\mathbb{Z}$-modules and $\mathbb{Z}$-morphisms of the form

$$0 \longrightarrow \mathbb{Z} \longrightarrow A \longrightarrow Y \longrightarrow 0$$

that does not split.

4.29 An R-morphism $f : M \to N$ is said to be *regular* if there is an R-morphism $g : M \to N$ such that $f \circ g \circ f = f$.

By considering the combined canonical sequences

$$\mathrm{Ker}\, f \xrightarrow{i_1} M \xrightarrow{\natural_1} M/\mathrm{Ker}\, f \simeq \mathrm{Im}\, f \xrightarrow{i_2} N \xrightarrow{\natural_2} N/\mathrm{Im}\, f$$

and examining when each splits, prove that $f : M \to N$ is regular if and only if $\mathrm{Ker}\, f$ is a direct summand of M and $\mathrm{Im}\, f$ is a direct summand of N.

4.30 In the diagram of R-modules and R-morphisms

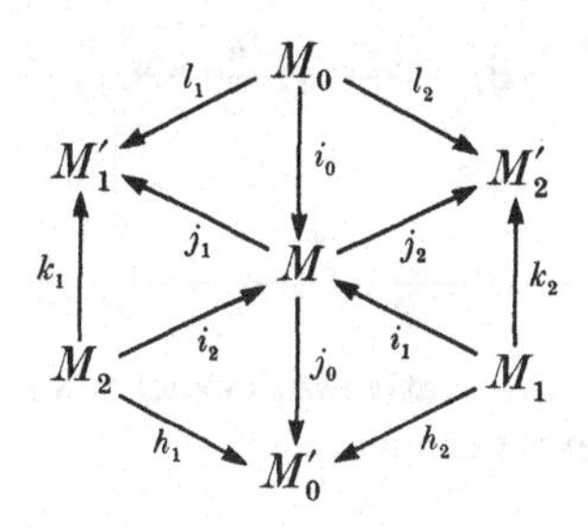

it is given that

(1) the diagram is commutative;
(2) each sequence $M_t \xrightarrow{i_t} M \xrightarrow{j_t} M'_t$ is exact;
(3) k_1 and k_2 are isomorphisms.

Prove that $\operatorname{Ker} j_1 \cap \operatorname{Ker} j_2 = 0$. If, for every $x \in M$, the element $\bar{x}$ is defined by

$$\bar{x} = (i_1 \circ k_2^{-1} \circ j_2)(x) + (i_2 \circ k_1^{-1} \circ j_1)(x),$$

deduce that $\bar{x} = x$. Hence show that $M = \operatorname{Im} i_1 \oplus \operatorname{Im} i_2$ and that

$$h_1 \circ k_1^{-1} \circ \ell_1 + h_2 \circ k_2^{-1} \circ \ell_2 = 0.$$

4.31 Let $(M_i)_{i\in I}$ and $(N_i)_{i\in I}$ be families of R-modules. If, for every $i \in I$, $f_i : M_i \to N_i$ is an R-morphism, define the *direct sum of the family* $(f_i)_{i\in I}$ to be the R-morphism $f : \bigoplus_{i\in I} M_i \to \bigoplus_{i\in I} N_i$ given by $f\big((m_i)_{i\in I}\big) = \big(f_i(m_i)\big)_{i\in I}$. Determine $\operatorname{Im} f$ and $\operatorname{Ker} f$.

If $(L_i)_{i\in I}$ is also a family of R-modules and if, for every $i \in I$, $g_i : L_i \to M_i$ is an R-morphism, let g be the direct sum of the family $(g_i)_{i\in I}$. Prove that

$$\bigoplus_{i\in I} L_i \xrightarrow{g} \bigoplus_{i\in I} M_i \xrightarrow{f} \bigoplus_{i\in I} N_i$$

is exact if and only if

$$L_i \xrightarrow{g_i} M_i \xrightarrow{f_i} N_i$$

is exact for every $i \in I$.

4.32 Let R be a commutative ring with a 1. Suppose that M is an R-module and that $A_1, \ldots, A_n$ are submodules of M with $M = \bigoplus_{i=1}^{n} A_i$. For $j = 1, \ldots, n$ let L_j be the set of R-morphisms $f : M \to N$ such that $\bigoplus_{i\neq j} A_i \subseteq \operatorname{Ker} f$. Prove that L_j is an R-module and that $L_j \simeq \operatorname{Mor}_R(A_j, N)$.

4.33 The diagram of R–modules and R–morphisms

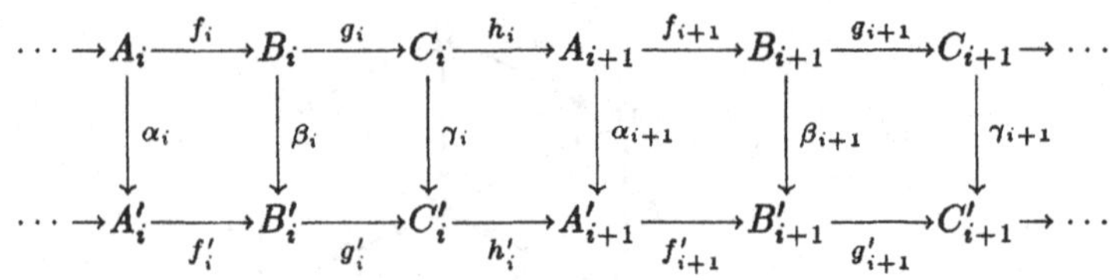

is given to be commutative with exact rows. If each γ_i is an isomorphism, establish the exact sequence

$$\cdots \longrightarrow A_i \xrightarrow{\varphi_i} A'_i \oplus B_i \xrightarrow{\vartheta_i} B'_i \xrightarrow{h_i\gamma_i^{-1}g'_i} A_{i+1} \longrightarrow \cdots$$

where φ_i is described by

$$\varphi_i : a_i \mapsto \big(\alpha_i(a_i), f_i(a_i)\big)$$

and ϑ_i is described by

$$\vartheta_i : (a'_i, b_i) \mapsto f'_i(a'_i) - \beta_i(b_i).$$

4.34 Give an example to show that a submodule of a free module need not be free.

4.35 Let $f : M \to M$ be an R–morphism. Prove that if f is a monomorphism then f is not a left zero divisor in the ring $\mathrm{End}_R(M)$. If M is free, establish the converse.

4.36 Let $f : M \to M$ be an R–morphism. Prove that if f is an epimorphism then f is not a right zero divisor in the ring $\mathrm{End}_R(M)$. Give an example of a free $\mathbb{Z}$–module M and $f \in \mathrm{End}_{\mathbb{Z}}(M)$ such that f is neither a right zero divisor nor an epimorphism.

4.37 If $(P_i)_{i\in I}$ is a family of projective modules prove that $\bigoplus_{i\in I} P_i$ is also projective.

4.38 For R–modules M and N, let $P(M, N)$ be the set of those R–morphisms $f : M \to N$ that 'factor through projectives' in the sense that there is a commutative diagram

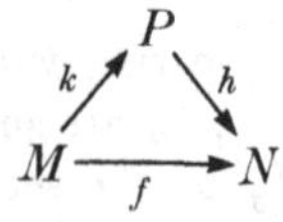

in which P is projective. Prove that $P(M, N)$ is a subgroup of the group $\mathrm{Mor}_R(M, N)$.

4.39 Show that every diagram of R–modules and R–morphisms of the form

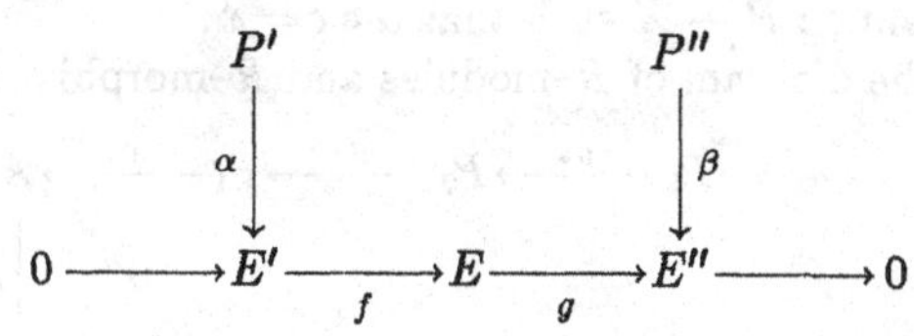

in which the row is exact and P', P'' are projective can be extended to a commutative diagram

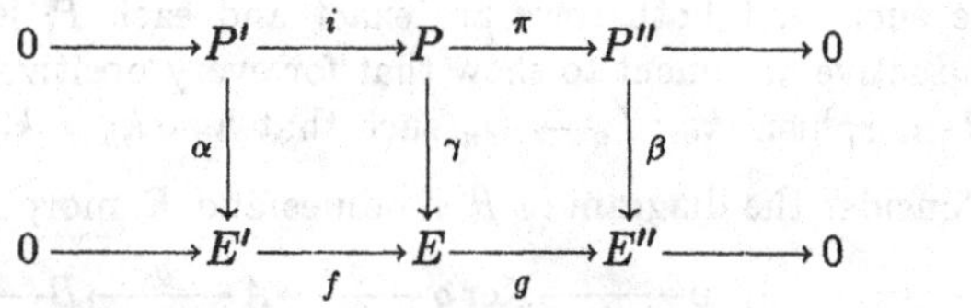

in which the top row is also exact and P is also projective.

4.40 Let n be an integer greater than 1. For every divisor r of n consider the ideal $r(\mathbb{Z}/n\mathbb{Z})$ of the ring $\mathbb{Z}/n\mathbb{Z}$. Show how to construct an exact sequence

$$0 \longrightarrow \frac{n}{r}(\mathbb{Z}/n\mathbb{Z}) \longrightarrow \mathbb{Z}/n\mathbb{Z} \longrightarrow r(\mathbb{Z}/n\mathbb{Z}) \longrightarrow 0.$$

Prove that the following statements are equivalent :

(1) the above sequence splits;
(2) h.c.f.$\{r, n/r\} = 1$;
(3) the $\mathbb{Z}/n\mathbb{Z}$–module $r(\mathbb{Z}/n\mathbb{Z})$ is projective.

Hence provide an example of a projective module that is not free.

4.41 Let R be a commutative ring with a 1, and let X, Y be R–modules with X projective. If A, B are submodules of X, Y respectively, prove that

$$\Delta_{A,B} = \{f \in \mathrm{Mor}_R(X,Y) \mid f^{\rightarrow}(A) \subseteq B\}$$

is a submodule of the R–module $\mathrm{Mor}_R(X,Y)$.
Prove that $\mathrm{Mor}_R(X/A, Y/B) \simeq \Delta_{A,B}/\Delta_{X,B}$.

4.42 Suppose that P is a projective R–module and that there is given a diagram of R–modules and R–morphisms of the form

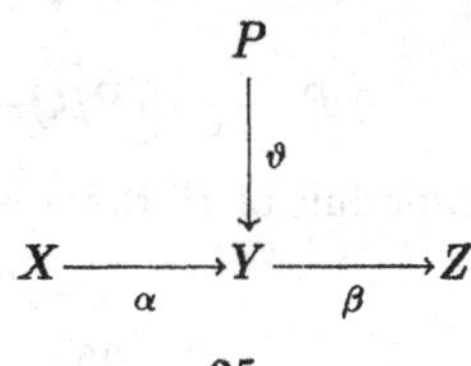

in which the row is exact and $\beta \circ \vartheta = 0$. Prove that there is an R–morphism $\varsigma : P \to X$ such that $\alpha \circ \varsigma = \vartheta$.

Let the diagram of R–modules and R–morphisms

$$\begin{array}{ccccccccc} \cdots \longrightarrow P_3 & \xrightarrow{g_3} & P_2 & \xrightarrow{g_2} & P_1 & \xrightarrow{g_1} & A & & \\ & & & & & & \downarrow{\scriptstyle k_0} & & \\ \cdots \longrightarrow Q_3 & \xrightarrow[h_3]{} & Q_2 & \xrightarrow[h_2]{} & Q_1 & \xrightarrow[h_1]{} & B & \longrightarrow & 0 \end{array}$$

be such that both rows are exact and each P_i is projective. Use an inductive argument to show that for every positive integer n there is an R–morphism $k_n : P_n \to Q_n$ such that $h_n \circ k_n = k_{n-1} \circ g_n$.

4.43 Consider the diagram of R–modules and R–morphisms

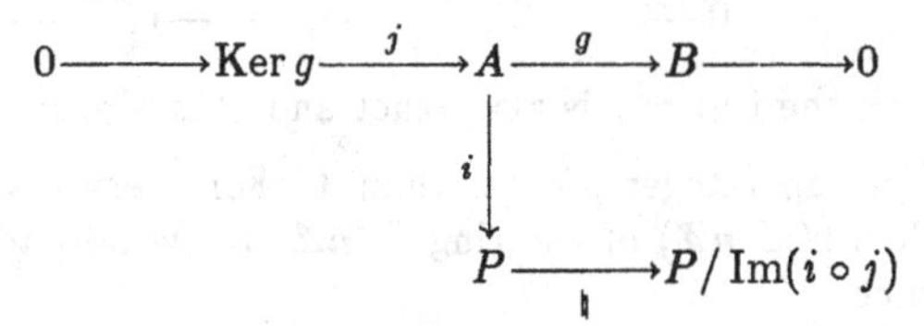

in which the top row is exact, j is the canonical inclusion, i is a monomorphism, and $\natural$ is the canonical epimorphism. Prove that there is a unique R–morphism $\vartheta : B \to P/\operatorname{Im}(i \circ j)$ such that $\vartheta \circ g = \natural \circ i$. Show also that ϑ is a monomorphism.

An R–module P is said to be ***quasi-projective*** if, for every diagram

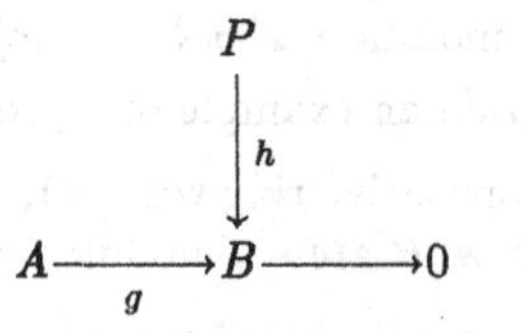

in which the row is exact and A is a submodule of P, there is a unique R–morphism $\varsigma : P \to A$ such that $g \circ \varsigma = h$. Prove that P is quasi-projective if and only if, for every diagram

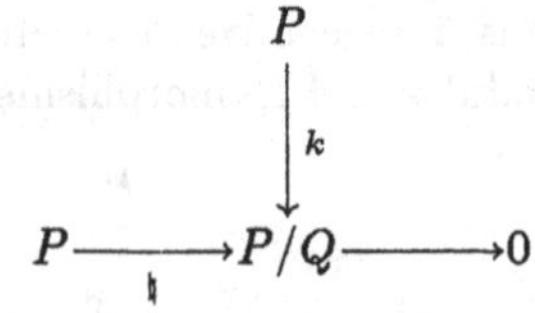

in which Q is a submodule of P, there is a unique R–morphism $\pi : P \to P$ such that $\natural \circ \pi = k$.

Solutions to Chapter 1

1.1 (a) To say that R/I is an integral domain is equivalent to saying that it has no zero divisors, which is equivalent to saying that, for all $x, y \in R$, if $(x+I)(y+I) = 0+I$ then either $x+I = 0+I$ or $y+I = 0+I$, which is equivalent to the condition $xy \in I$ implies $x \in I$ or $y \in I$, i.e. to the condition that I is prime.

(b) Suppose that I is maximal and let $x+I$ be a non-zero element of R/I. Then $x \notin I$ and so $I+(x)$ is an ideal of R that properly contains I. Hence $I+(x) = R$ and so $i+rx = 1$ for some $i \in I$ and $r \in R$. Passing to quotients (i.e. applying the natural morphism associated with I to this equation), we obtain $(r+I)(x+I) = 1+I$ which shows that $r+I$ is an inverse of $x+I$. Thus R/I is a field. Conversely, if R/I is a field and J is an ideal of R with $I \subset J$, let $b \in J \setminus I$. Since $b+I \neq 0+I$ it follows that for every $r \in R$ there exists $x \in R$ such that $(b+I)(x+I) = r+I$. It follows that $bx - r = i \in I$ and hence that $r = bx - i \in J$ and consequently $J = R$.

Since every field is an integral domain, it follows from the above that every maximal ideal is prime.

Clearly, the ideal (0) of $\mathbb{Z}$ is prime but not maximal.

If every non-zero element of R/M is invertible then R/M has no proper non-zero ideals. For, if A were such and $a \in A$ for some $a \neq 0$ then a^{-1} exists and $a^{-1}a \in A$ gives the contradiction $A = R/M$. Hence R has no proper ideals that properly contain M.

Now

$$\text{Mat}_{2\times 2}(\mathbb{Z})/M \simeq \text{Mat}_{2\times 2}(\mathbb{Z}_2)$$

and so, since $\text{Mat}_{2\times 2}(\mathbb{Z}_2)$ contains no proper ideals, M is maximal in

$\mathrm{Mat}_{2\times 2}(\mathbb{Z})$. However, $\begin{bmatrix} 0 & 1 \\ 0 & 0 \end{bmatrix} + M$ is not invertible in $\mathrm{Mat}_{2\times 2}(\mathbb{Z})/M$ since this would imply

$$\begin{bmatrix} 0 & 1 \\ 0 & 0 \end{bmatrix}\begin{bmatrix} a & b \\ c & d \end{bmatrix} = \begin{bmatrix} 1+2x & 2y \\ 2z & 1+2t \end{bmatrix}$$

which is impossible.

1.2 Suppose that $(8) \subseteq I \subseteq 4\mathbb{Z}$. If $I \neq (8)$ then there exists $t = 4n \in I$ with n odd, say $n = 2m+1$. Then $t = 8m+4$ which gives $t + (8) = 4 + (8)$ in $\mathbb{Z}/(8)$. Hence $I = 4\mathbb{Z}$ and $4\mathbb{Z}/(8)$ has no non-trivial ideals. Clearly, $4\mathbb{Z}/(8)$ is not a field since if $x = 4 + (8)$ then we have $x^2 = 0$ in $\mathbb{Z}/(8)$.
The reason this can happen is that $4\mathbb{Z}$ has no identity element.

1.3 (a) Since every element is idempotent, we have

$$x + x = (x+x)^2 = x^2 + x^2 + x^2 + x^2 = x + x + x + x$$

whence $x + x = 0$ and the characteristic is 2.
(b) We have that

$$x + y = (x+y)^2 = x^2 + xy + yx + y^2 = x + xy + yx + y$$

so $0 = xy + yx$. Using the fact that $y + y = 0$ implies that $-y = y$, we then obtain $xy = -yx = (-y)x = yx$.
(1) $\Rightarrow$ (2) : If I is prime then A/I is an integral domain, so has no zero divisors. Now in A we have

$$xy(x+y) = xyx + xy^2 = x^2y + xy^2 = xy + xy = 0.$$

On passing to quotients modulo I this identity still holds, say $\bar{x}\bar{y}(\bar{x}+\bar{y}) = \bar{0}$. If then $\bar{x} \neq \bar{0}$ and $\bar{y} \neq \bar{0}$ then we must have $\bar{x} + \bar{y} = \bar{0}$ whence $\bar{x} = -\bar{y} = \bar{y}$. Thus A/I consists of two elements and so $A/I \simeq \mathbb{Z}/2\mathbb{Z}$.
(2) $\Rightarrow$ (3) : This follows by question 1.1.
(3) $\Rightarrow$ (1) : This follows by question 1.1.

1.4 $J = X\,F[X,Y]$ so J is principal. Now we have that $F[X,Y]/J \simeq F[Y]$ which is an integral domain. Hence J is a prime ideal of $F[X,Y]$. However, since $F[Y]$ is not a field it follows that J is not a maximal ideal of $F[X,Y]$. In fact,

$$J \subset (J, Y^2) \subset F[X,Y].$$

1.5 (a) Since $A \subseteq r(A)$ for every A we have

$$I + J \subseteq r(I) + r(J) \subseteq r[r(I) + r(J)].$$

Now let $x \in r[r(I) + r(J)]$. Then $x^n \in r(I) + r(J)$ for some $n \in \mathbb{N}$ and so $x^n = y + z$ where $y^{m_1} \in I$ and $z^{m_2} \in J$ for some $m_1, m_2 \in \mathbb{N}$. Hence $x^{n(m_1+m_2)} = \sum y^\mu z^\nu$ where $\mu + \nu = m_1 + m_2$ and either $\mu \geq m_1$ or $\nu \geq m_2$. But if $\mu \geq m_1$ we have $y^\mu z^\nu \in I$, while if $\nu \geq m_2$ we have $y^\mu z^\nu \in J$. Hence $x^{n(m_1+m_2)} \in I + J$ so $x \in r(I + J)$.

(b) If $I^n \subseteq J$ let $x \in r(I)$. Then $x^m \in I$ for some m. But then $(x^m)^n \in J$ since $I^n \subseteq J$, and consequently $x \in r(J)$.

The set of nilpotent elements of R/I is $r(I/I) = r(I)/I$. Hence R/I has no non-zero nilpotent elements if and only if $I = r(I)$.

If P is prime and $x^n \in P$ then $x \in P$. Thus if $x \in r(P)$ we have $x \in P$, and so $r(P) = P$.

1.6 (a) If $x \in I : K$ then $xk \in I$ for all $k \in K$ and so $xk \in J$ for all $k \in K$, whence $x \in J : K$.

If now $x \in K : J$ then $xa \in K$ for all $a \in J$ and so $xa \in K$ for all $a \in I$, whence $x \in K : I$.

(b) Observe first the general result that

$$(I : J) : K = I : JK.$$

In fact $JK[(I : J) : K] \subseteq J(I : J) \subseteq I$ and so it follows that $(I : J) : K \subseteq I : JK$. Also, $JK(I : JK) \subseteq I$ gives $K(I : JK) \subseteq I : J$ whence $I : JK \subseteq (I : J) : K$. Now (b) follows on taking $K = J^n$.

(c) If $J \subseteq I$ and $a \in R$ then $aJ \subseteq J \subseteq I$ gives $a \in I : J$. Conversely, if $I : J = R$ then in particular we have $1 \in I : J$ whence $J = 1J \subseteq I$.

(d) $I : (I + J) = (I : I) \cap (I : J) = R \cap (I : J) = I : J$.

1.7 Clearly (p^n) is a primary ideal of $\mathbb{Z}$ when p is prime. Conversely, suppose that $m = pq$ with p, q coprime. Then $pq \in (m)$ and $p \notin (m)$ yet $q^n \notin (m)$ for any $n \geq 1$. Therefore (m) is not primary.

It is straightforward to check that $(4, X)$ is a primary ideal. For, if $fg \in (4, X)$ with $f \notin (4, X)$ then any term of g not divisible by 4 or X must be divisible by 2. Hence $g^2 \in (4, X)$, so $(4, X)$ is primary.

If $(4, X)$ is the power of a prime ideal P then P must contain $(4, X)$. But $\mathbb{Z}[X]/(4, X) \simeq \mathbb{Z}_4$ and so the only possibility is $P = (2, X)$ which corresponds to $2\mathbb{Z}_4$, the only proper non-zero ideal of $\mathbb{Z}_4$. However, $(2, X)^n \neq (4, X)$ for any $n \geq 1$.

Now $R/P \simeq \mathbb{Z}$ so P is a prime ideal of R (since $\mathbb{Z}$ is an integral domain). However, $3X^3 \in P^2$ since $3X \in P$ and $X^2 \in P$. But $X^3 \notin P$, so if P^2 were primary we would have $3^n \in P^2$ for some $n \geq 1$. Since every element of P^2 has constant term 0, this is a contradiction.

1.8 Clearly $(Y^2Z^2, XYZ) \subseteq (Y) \cap (Z) \cap (X,Y)^2 \cap (X,Z)^2$. To obtain the reverse inclusion, suppose that $f(X,Y,Z)$ is in the right hand side. Then since $f \in (Y) \cap (Z)$ we have $f(X,Y,Z) = YZ\,g(X,Y,Z)$. But we also have $f \in (X,Y)^2$ so

$$g(X,Y,Z) = Y\,a(X,Y,Z) + X\,b(X,Y,Z).$$

Now $YZX\,b(X,Y,Z) \in (X,Z)^2$ so we require $Y^2Z\,a(X,Y,Z) \in (X,Z)^2$, i.e.

$$a(X,Y,Z) = Z\,h(X,Y,Z) + X\,k(X,Y,Z).$$

Then $f \in (Y^2Z^2, XYZ)$ as required.

It is not true that $(Y^2Z^2, XYZ) = (Y)(Z)(X,Y)^2(X,Z)^2$. For, every polynomial in the right hand side has degree at least 6 and so in particular XYZ cannot belong to the right hand side.

1.9 $F_I \neq \emptyset$ since clearly $I \in F_I$. Let $\{J_\beta \mid \beta \in B\}$ be a totally ordered subset of F_I. Then $\bigcup_{\beta \in B} J_\beta \neq A$, for otherwise we would have $1 \in J_\beta$ for some β whence the contradiction

$$A = A1 \subseteq AJ_\beta \subseteq J_\beta \subset A.$$

Suppose now that $x, y \in \bigcup_{\beta \in B} J_\beta$. Then there exist $\alpha, \gamma \in B$ such that $x \in J_\alpha$ and $y \in J_\gamma$. But either $J_\alpha \subseteq J_\gamma$ or $J_\gamma \subseteq J_\alpha$. Suppose, without loss, that $J_\alpha \subseteq J_\gamma$. Then $x - y \in J_\gamma \subseteq \bigcup_{\beta \in B} J_\beta$. Similarly $xa, ax \in J_\gamma$ for every $a \in A$. Thus we see that $\bigcup_{\beta \in B} J_\beta$ is an ideal of A that is distinct from A. Thus F_I is inductively ordered.

Applying Zorn's axiom, we obtain a maximal element M of F_I. It is clear that M is also a maximal ideal of A with $I \subseteq M$.

Let $I \in I(A)$ with $I \neq A$. Then no element of I is invertible. For, suppose that $x \in I$ is invertible. Then $1 = x^{-1}x \in I$ gives $A = I$, a contradiction. It follows that if $a \in A$ is invertible then a does not belong to any proper ideal, hence to any maximal ideal. Conversely, if $a \in A$ does not belong to any maximal ideal then, by the first part of the question, a does not belong to any proper ideal. Now $Aa = aA$ is an ideal (by commutativity) of A with $a = 1a \in Aa$. We deduce, therefore, that $Aa = A$ whence there exists $x \in A$ such that $xa = 1$. Thus a is invertible.

1.10 $\mathrm{Mat}_{n\times n}(F)$ is a finite-dimensional vector space over F. A left or right ideal of this is a subspace. Thus, if L_1, L_2 are left ideals with $L_1 \subset L_2$ then $\dim L_1 < \dim L_2$. It follows that every chain of left ideals is finite and so both chain conditions are satisfied. The same holds for right ideals.

Let I be a non-zero proper right ideal of R. Suppose that I contains the non-zero elements

$$\begin{bmatrix} c_1 & 0 \\ c_2 & 0 \end{bmatrix}, \qquad \begin{bmatrix} d_1 & 0 \\ d_2 & 0 \end{bmatrix}.$$

Suppose in fact that $c_1 \neq 0$ and $d_1 \neq 0$. Then

$$\begin{bmatrix} c_1 & 0 \\ c_2 & 0 \end{bmatrix}\begin{bmatrix} c_1^{-1} & 0 \\ 0 & 0 \end{bmatrix} = \begin{bmatrix} 1 & 0 \\ c_1^{-1}c_2 & 0 \end{bmatrix} \in I$$

and similarly

$$\begin{bmatrix} 1 & 0 \\ d_1^{-1}d_2 & 0 \end{bmatrix} \in I.$$

But then I contains

$$\begin{bmatrix} 1 & 0 \\ c_1^{-1}c_2 & 0 \end{bmatrix} - \begin{bmatrix} 1 & 0 \\ d_1^{-1}d_2 & 0 \end{bmatrix} = \begin{bmatrix} 0 & 0 \\ d & 0 \end{bmatrix}$$

where $d = c_1^{-1}c_2 - d_1^{-1}d_2$. Now if $d \neq 0$ then we have

$$\begin{bmatrix} 0 & 0 \\ 1 & 0 \end{bmatrix} = \begin{bmatrix} 0 & 0 \\ d & 0 \end{bmatrix}\begin{bmatrix} d^{-1} & 0 \\ 0 & 0 \end{bmatrix} \in I$$

whence

$$\begin{bmatrix} 0 & 0 \\ c_2 & 0 \end{bmatrix} = \begin{bmatrix} 0 & 0 \\ 1 & 0 \end{bmatrix}\begin{bmatrix} c_2 & 0 \\ c_1 & 0 \end{bmatrix} \in I$$

and therefore also

$$\begin{bmatrix} c_1 & 0 \\ 0 & 0 \end{bmatrix} = \begin{bmatrix} c_1 & 0 \\ c_2 & 0 \end{bmatrix} - \begin{bmatrix} 0 & 0 \\ c_2 & 0 \end{bmatrix} \in I.$$

Consequently,

$$\begin{bmatrix} 1 & 0 \\ 0 & 0 \end{bmatrix} = \begin{bmatrix} c_1 & 0 \\ 0 & 0 \end{bmatrix}\begin{bmatrix} c_1^{-1} & 0 \\ 0 & 0 \end{bmatrix} \in I.$$

It now readily follows that $I = R$, a contradiction. Hence we must have $d = 0$ and so $d_1c_2 = c_1d_2$ which gives

$$\begin{bmatrix} c_1 & 0 \\ c_2 & 0 \end{bmatrix} = \alpha \begin{bmatrix} d_1 & 0 \\ d_2 & 0 \end{bmatrix}$$

for some $\alpha \in F$. If $c_1 = 0$ then $d_1 = 0$, otherwise $I = R$. Hence the result follows.

No two proper non-zero right ideals can contain each other and so R has both chain conditions on right ideals.

If S is an additive subgroup of $\mathbb{R}$ then $S\begin{bmatrix} 0 & 0 \\ 1 & 0 \end{bmatrix}$ is a left ideal of R. Then

$$2\mathbb{Z}\begin{bmatrix} 0 & 0 \\ 1 & 0 \end{bmatrix} \supset 4\mathbb{Z}\begin{bmatrix} 0 & 0 \\ 1 & 0 \end{bmatrix} \supset 8\mathbb{Z}\begin{bmatrix} 0 & 0 \\ 1 & 0 \end{bmatrix} \supset \dots$$

is an infinite descending chain of left ideals.

Taking $S_i = \{t/i^n \mid t, n \in \mathbb{Z}\}$ we obtain the infinite ascending chain of left ideals

$$S_2\begin{bmatrix} 0 & 0 \\ 1 & 0 \end{bmatrix} \subset S_4\begin{bmatrix} 0 & 0 \\ 1 & 0 \end{bmatrix} \subset S_8\begin{bmatrix} 0 & 0 \\ 1 & 0 \end{bmatrix} \subset \dots .$$

1.11 Suppose that $R = \{r + xr \mid r \in R\}$. Then $x = r + xr$ for some $r \in R$ and so $x + (-r) + x(-r) = 0$, which shows that x is right quasi-regular. Conversely, if $x + y + xy = 0$ for some $y \in R$ then $x = -y + x(-y) \in \{r + xr \mid r \in R\}$. Since this is a right ideal of R, it then contains xr and hence also $r + xr - xr = r$ so coincides with R.

Suppose now that x belongs to every maximal right ideal of R. If x is not right quasi-regular then $A = \{r+xr \mid r \in R\} \neq R$. Let M be a right ideal of R that is maximal with respect to containing the set A but not x. Then M is a maximal right ideal of R and so $x \in M$, a contradiction. Hence x is right quasi-regular, so every element in the intersection of the maximal right ideals of R is right quasi-regular, which shows that xr is right quasi-regular for every $r \in R$.

If M is a maximal right ideal of R with $x \notin M$ then $\{M+xr \mid r \in R\}$ is a right ideal of R with $M \subset \{M+xr \mid r \in R\}$ so $\{M+xr \mid r \in R\} = R$ whence we have $-1 = m + xr$ for some $m \in M$ and $r \in R$.

Suppose now tht xr is right quasi-regular for every $r \in R$ and that $x \notin M$ for some maximal right ideal M. Then $-1 = m + xr$ and xr is right quasi-regular, so there exists $z \in R$ with $xr + z + xrz = 0$. But $-z = mz + xrz$ gives $mz = xr$. Hence $xr \in M$ and so $-1 = m + xr \in M$ which shows that $M = R$, a contradiction. Hence the result follows.

1.12 The result is true (by hypothesis) for $j = i$. Suppose that $j > i$ and that $R^{j-1} = S^{j-1} + R^j$. Then

$$\begin{aligned} R^j = R^{j-1}R &= (S^{j-1} + R^j)R \\ &\subseteq SR^{j-1} + R^{j+1} \\ &= S(S^{j-1} + R^j) + R^{j+1} \\ &\subseteq S^j + R^{j+1}. \end{aligned}$$

Clearly, $S^j + R^{j+1} \subseteq R^j$ and so $R^j = S^j + R^{j+1}$.

Using induction and the above, we have

$$\begin{aligned} R^j &= S^j + R^{j+k-1} \\ &= S^j + S^{j+k-1} + R^{j+k} \\ &= S^j + R^{j+k}. \end{aligned}$$

Suppose now that R is nilpotent, say $R^m = \{0\}$.

(a) $R^i = S^i + R^{i+1}$ implies $R^j = S^j + R^{j+k}$ for $j \geq i$ and any k. Thus $R^j = S^j + R^{j+m} = S^j$ for all $j \geq i$.

(b) Let $R = \langle a \rangle + R^2$. Then by (a) we have $R^j = \langle a \rangle^j$ for all $j \geq 1$. Take $j = 1$ to obtain $R = \langle a \rangle$.

(c) If $R^2 \not\subseteq M$ then $R = M + R^2$ since M is maximal. Hence, again by (a), $R = M$ which is impossible. Hence $R^2 \subseteq M$.

(d) Since $R^2 \subseteq M$ the quotient R/M has zero multiplication, so every subgroup of the additive group M is an ideal of R. Hence R/M has a prime number of elements.

1.13 Let A be a subring of a nilring R. If $a \in A$ then $a \in R$ so a is nilpotent. Hence A is a nilring. Also, if $b \in R/I$ then $b = x + I$ for some $x \in R$, and since $x^n = 0$ for some $n \geq 1$ we have $b^n = I$. Thus R/I is a nilring.

Suppose now that A and B/A are nil. Let $x \in B$. Since $(x + A)^n = A$ for some $n \geq 1$ we have $x^n \in A$ whence $(x^n)^m = 0$ since A is nil. Thus x is nilpotent and so B is nil.

If B is nil then so is $B/(A \cap B) \simeq (A + B)/A$. But A is nil. Hence so is $A + B$.

In $\mathbb{R}_p$ we have $(a_1, a_2, a_3, \ldots)^n = 0$ if and only if $a_i^n = 0$ for $i \geq 1$, which is the case if and only if every a_i is divisible by p. Hence, since there are only finitely many non-zero a_i, we have

$$(a_1, a_2, a_3, \ldots)^n = 0 \iff a_i \in p\mathbb{Z}_{p^i}.$$

Now it is easy to see that the set N of nilpotent elements is a two-sided ideal of R_p. However, N is not nilpotent. For if it were then we would have $N^m = \{0\}$ for some m, giving $(a_1, a_2, \ldots)^m = 0$ for all $(a_1, a_2, \ldots) \in N$. However,

$$(\underbrace{0,0,\ldots,0,p}_{m+1},0,\ldots)^m \neq (0,0,\ldots).$$

1.14 (a) $J \subseteq A_1$ since J is the smallest ideal of A_2 containing I. Thus $JI \subseteq I$ and $IJ \subseteq I$. Hence

$$\begin{aligned} J^3 &= J(I + A_2 I + IA_2 + A_2 I A_2)J \\ &\subseteq I + JI + IJ + JIJ \\ &\subseteq I. \end{aligned}$$

(b) If $I = A_0 \subset A_1 \subset \cdots \subset A_n = R$, call I an n–step subideal. Let $\bar{I}$ be the smallest ideal of R containing I. We show that $\bar{I}^{3n} \subseteq I$ by induction on n.

If J is the smallest ideal of A_2 containing I then

$$J \subset A_2 \subset A_3 \cdots \subset A_n = R$$

and J is an $(n-1)$–step subideal. By the induction hypothesis, $\bar{J}^{3n-3} \subseteq J$ where $\bar{J}$ is the smallest ideal of R containing J. But since $I \subseteq A_1 \subseteq A_2$ and J is the smallest ideal of A_2 containing I we have $J^3 \subseteq I$ by (a). Consequently, $\bar{J}^{3n} \subseteq J.J^3 \subseteq JI \subseteq I$. However, $I \subseteq J \subseteq \bar{I}$ so $\bar{J} = \bar{I}$ and then $\bar{I}^{3n} \subseteq I$.

(c) If $I^m = \{0\}$ we have $(\bar{I}^{3n})^m = \{0\}$ and $\bar{I}$ is nilpotent.

1.15 Suppose that Q is semiprime and that $A^n \subseteq Q$. We show that $A \subseteq Q$. The result clearly holds for $n = 2$. Suppose by way of induction that if $A^m \subseteq Q$ with $m < n$ then $A \subseteq Q$. If n is even then $A^{\frac{1}{2}n} A^{\frac{1}{2}n} \subseteq Q$ implies $A^{\frac{1}{2}n} \subseteq Q$ whence $A \subseteq Q$ by the inductive hypothesis. If n is odd then $A^{n+1} \subseteq Q$ and $n+1$ is even, whence again $A \subseteq Q$.

Let $\natural : R \to R/Q$ be the natural morphism. Suppose that Q is semiprime and that X is nilpotent in R/Q, say $X^n = 0$. Then, denoting the inverse image of a subset T of R/Q under $\natural$ by $\overline{T}$, we have $\overline{X}^n \subseteq \overline{X^n} = Q$. Since Q is semiprime it follows that $\overline{X} \subseteq Q$ and so $X = 0$.

Conversely, suppose that R/Q contains no non-zero nilpotent ideals. Let A be an ideal with $A^2 \subseteq Q$. Then $[\natural(A)]^2 = \natural(A^2) = 0$ whence $\natural(A) = 0$ and $A \subseteq Q$.

Solutions to Chapter 2

2.1 (a) $x|y \iff y = xr \iff y \in (x) \iff (y) \subseteq (x)$.

(b) If x and y are associates then $(x) = (y)$ follows from (a). Conversely, if $(x) = (y)$ then there exist $r, s \in R$ such that $x = ry, y = sx$ whence $x = rsx$ giving $rs = 1$ so that r, s are units and x, y are associates.

(c) Using (b) we have

$$u \text{ is a unit} \iff u \sim 1 \iff (u) = (1) = R.$$

(d) Suppose that $x = yr$ where y, r are not units. Then certainly $(x) \subseteq (y) \subseteq R$. Now $(x) \neq (y)$ since otherwise x, y are associates (by (b)) and then r would be a unit; and $(y) \neq R$ since otherwise y would be a unit (by (c)). The converse is clear.

(e) If (x) is not maximal among the principal ideals then for some y we have $(x) \subset (y) \subset R$. By (d), y is then a proper factor of x and so x is not irreducible. Conversely, if x is not irreducible then it has a proper factor y and the result follows from (d).

2.2 (a) True. If $a|b$ implies $b|a$ then every pair of non-zero elements are associates (for then $x \sim xy \sim y$). Hence $x \sim 1$ for every $x \in R^\star$ and so R is a field.

(b) True, as observed in the proof of (a).

(c) $a = bu$ and $c = dv$ give $ac = bd.uv$ and so $ac \sim bd$.

(d) False. We have $1 \sim 1$ and $1 + 2i \sim -2 + i$ in $\mathbb{Z}[i]$ but $1 + (1 + 2i)$ is not an associate of $(-2 + i) + 1$.

(e) True. If every element of $R^\star$ is a unit or a prime then every element of $R^\star$ must be a unit. For, suppose that $p \in R^\star$ is prime and consider p^2. Clearly, p^2 is not prime, and it cannot be a unit since $p^2 \in (p) \neq R$. Thus every element of $R^\star$ is a unit and so R is a field.

2.3 In $\mathbb{Z}$, 5 is irreducible since it is prime.

In $\mathbb{Z}[X]$, 5 is irreducible since $5 = f(X)g(X)$ implies that $\deg f(X) = \deg g(X) = 0$ so $f(X)$ and $g(X)$ are constant polynomials. Then one of f, g is ± 1.

In $\mathbb{Z}[i]$ we have $5 = (2+i)(2-i)$ where neither $2+i$ nor $2-i$ is a unit (since $\ell(2+i) = \ell(2-i) = 5$), so 5 is not irreducible.

In $\mathbb{Z}[\sqrt{-2}]$, 5 is irreducible. In fact, if $5 = \alpha\beta$ then $25 = \ell(5) = \ell(\alpha)\ell(\beta)$. But if $\alpha = a + b\sqrt{-2}$ then $\ell(\alpha) = a^2 + 2b^2 \neq \pm 5$.

2.4 We have that $\ell(43i - 19) = 2210 = 2 \cdot 5 \cdot 13 \cdot 17$. Consider now $\alpha = a + ib$. We have that α is of length

(1) 2 if and only if $a^2 + b^2 = 2$, which is the case if and only if α is an associate of $1 + i$;

(2) 5 if and only if $a^2 + b^2 = 5$, which is the case if and only if α is an associate of $2 + i$ or of $1 + 2i$;

(3) 13 if and only if $a^2 + b^2 = 13$, which is the case if and only if α is an associate of $3 + 2i$ or of $2 + 3i$;

(4) 17 if and only if $a^2 + b^2 = 17$, which is the case if and only if α is an associate of $4 + i$ or of $1 + 4i$.

It is now as easy matter to check that

$$43i - 19 = (2+3i)(1+i)(2+i)(4-i).$$

2.5 (a) $11 + 7i$ has length $170 = 2 \cdot 5 \cdot 17$. Now $1 + i$ has length 2, and

$$(11+7i)\frac{1-i}{2} = \tfrac{1}{2}(11 - 11i + 7i + 7) = 9 - 2i$$

(which has length $5 \cdot 17$) and so $11 + 7i = (1+i)(9-2i)$. Now $1 + 2i$ has length 5 and

$$(9-2i)\frac{1-2i}{5} = \tfrac{1}{5}(9 - 2i - 18i - 4) = 1 - 4i$$

(which has length 17) and so $9 - 2i = (1+2i)(1-4i)$. Thus

$$11 + 7i = (1+i)(1+2i)(1-4i).$$

(b) $4 + 7\sqrt{2} = \sqrt{2}(7 + 2\sqrt{2})$.

(c) $4 - \sqrt{-3}$ has length 19 which is prime so $4 - \sqrt{-3}$ is irreducible in $\mathbb{Z}[\sqrt{-3}]$.

2.6 Unique factorisation is not violated since

$$\begin{aligned} 6 = 2 \cdot 3 &= [(-2+\sqrt{6})(2+\sqrt{6})][(3-\sqrt{6})(3+\sqrt{6})] \\ &= [(-2+\sqrt{6})(3+\sqrt{6})][(2+\sqrt{6})(3-\sqrt{6})] \\ &= \sqrt{6} \cdot \sqrt{6}. \end{aligned}$$

In $\mathbb{Z}[\sqrt{10}]$ we have

$$6 = 2 \cdot 3 = (4+\sqrt{10})(4-\sqrt{10})$$

which gives two distinct factorisations of 6 into irreducibles.

2.7 In $\mathbb{Z}[\sqrt{-6}]$ the only units are ± 1 since $a+b\sqrt{-6}$ is a unit if and only if $\ell(a+b\sqrt{-6}) = a^2+6b^2 = \pm 1$, i.e. if $b=0$ and $a=\pm 1$. Consider now the decompositions

$$10 = 2 \cdot 5 = (2+\sqrt{-6})(2-\sqrt{-6}).$$

No element of $\mathbb{Z}[\sqrt{-6}]$ can have length 5 since $a^2+6b^2 = 5$ is clearly impossible. Hence $2+\sqrt{-6}$ and $2-\sqrt{-6}$ are irreducible since each has length 10 and any decomposition would involve an element of length 5. Similarly, 5 has length 25 and is irreducible. Also, since $a^2+6b^2 = 2$ is also impossible, no element can have length 2 and hence 2 (of length 4) is irreducible. We thus have two distinct factorisations of 10 into products of irreducibles and so $\mathbb{Z}[\sqrt{-6}]$ is not a unique factorisation domain.

(a) $2+\sqrt{-6}$ is irreducible but not prime, for it divides 10 but does not divide either 2 or 5.

(b) The greatest common divisor of 10 and $2(2+\sqrt{-6})$ does not exist. In fact, 2 and $2+\sqrt{-6}$ each divide 10 but $2(2+\sqrt{-6})$ does not divide 10 since $\ell[2(2+\sqrt{-6})] = 40$ which does not divide $\ell(10) = 100$.

(c) Since 5 and $2+\sqrt{-6}$ are irreducible their greatest common divisor exists and is 1. Suppose that $5\alpha + (2+\sqrt{-6})\beta = 1$. Then $10\alpha + 2(2+\sqrt{-6})\beta = 2$ which is impossible since $2+\sqrt{-6}$ divides the left hand side but does not divide the right hand side.

2.8 We have

$$8 = 2 \cdot 2 \cdot 2 = (1+\sqrt{-7})(1-\sqrt{-7}).$$

Now each of $2, 1+\sqrt{-7}, 1-\sqrt{-7}$ is irreducible. For these elements have lengths $4, 8, 8$ respectively so if any were reducible then there would exist $a+b\sqrt{-7}$ with $a^2+7b^2 = 2$ which is clearly impossible.

Now $8^k \in \mathbb{Z}[\sqrt{-7}]$ can be written as

$$\begin{aligned} 8^k &= (1-\sqrt{-7})^k(1+\sqrt{-7})^k && 2k \text{ irreducibles} \\ &= (1-\sqrt{-7})^{k-1}(1+\sqrt{-7})^{k-1}2^3 && 2k+1 \text{ irreducibles} \\ &= \dots \\ &= (1-\sqrt{-7})^{k-i}(1+\sqrt{-7})^{k-i}2^{3i} && 2k+i \text{ irreducibles} \\ &= \dots \\ &= 2^{3k} && 3k \text{ irreducibles} \end{aligned}$$

whence the result follows on taking $t = 2k$.

An element that is irreducible but not prime is 2; for 2 divides $8 = (1+\sqrt{-7})(1-\sqrt{-7})$ but divides neither of the factors.

Since 2 and $1+\sqrt{-7}$ are irreducible they have greatest common divisor 1. Suppose that $\gamma 2 + \delta(1+\sqrt{-7}) = 1$. Then on multiplying each side by $1-\sqrt{-7}$ we obtain

$$\gamma 2(1-\sqrt{-7}) + \delta 8 = 1 - \sqrt{-7}.$$

But 2 divides the left hand side and does not divide the right hand side, so we have the required contradiction.

2.9 Let $I = (r)$. Then given $\alpha \in \mathbb{Z}[i]$ we have $\alpha = \beta r + \gamma$ where $N(\gamma) < N(r)$. Since $\gamma/I = \alpha/I$ and there are only finitely many γ with $N(\gamma) < N(r)$ it follows that $\mathbb{Z}[i]/I$ is finite.

2.10 Let $\alpha = a + ib$ be a prime in $\mathbb{Z}[i]$. Suppose first that $\ell(\alpha)$ is even. Then $a^2 + b^2$ is even. Since a, b cannot both be even (otherwise we can extract a factor 2 and α would not be prime), it follows that both must be odd. But then

$$a + ib = [\tfrac{1}{2}(a+b) + \tfrac{1}{2}(b-a)i](1+i)$$

whence $\alpha \sim 1+i$ since $\frac{1}{2}(a+b) + \frac{1}{2}(b-a)i$ must be a unit (otherwise $a+ib$ would not be prime).

Now suppose that $\ell(\alpha)$ is odd. Then if a, b are both non-zero we have $\ell(\alpha) = a^2 + b^2$ is prime in $\mathbb{Z}$; for otherwise $(a+ib)(a-ib) = a^2 + b^2$ and a decomposition of a^2+b^2 into primes in $\mathbb{Z}$ gives a contradiction to $a+ib$ being prime. Now $a+ib$ is a factor of $p = a^2+b^2$ and $p \equiv 1 \bmod 4$, which is (c).

If $b = 0$ then $\alpha = a \in \mathbb{Z}$ and a must be prime. If $a \equiv 1 \bmod 4$ then $a = c^2 + d^2$ for some $c, d \in \mathbb{Z}$ so again case (c) occurs. If $a \equiv 3 \bmod 4$ then case (b) occurs.

If $a = 0$ then α is an associate of $b \in \mathbb{Z}$ and the same argument works.

Solutions to Chapter 2

2.11 (a) If $n < -1$, say $n = -p$, then Pell's equation is $a^2 + bp^2 = 1$. The only solutions are $a = \pm 1, b = 0$ and so the group of units is $\{-1, 1\}$.

(b) If $n > 1$ then Pell's equation is $a^2 - nb^2 = \pm 1$. If we write $\alpha = a + b\sqrt{n} \in \mathbb{R}$ and $\bar{\alpha} = a - b\sqrt{n} \in \mathbb{R}$ then this becomes $\alpha\bar{\alpha} = \pm 1$. It follows that if α is a unit then so also is α^k for every integer $k \geq 1$, for then we have $\alpha^k \bar{\alpha}^k = \pm 1$.

(c) By (b) we see that the group of units of $\mathbb{Z}[\sqrt{2}]$ contains the set $\{\pm(1+\sqrt{2})^k \mid k \geq 1\}$. Suppose now that u is a positive unit of $\mathbb{Z}[\sqrt{2}]$. Then for some $k \geq 1$ we have

$$(1+\sqrt{2})^k \leq u < (1+\sqrt{2})^{k+1}$$

whence we deduce that

$$1 \leq u(1+\sqrt{2})^{-k} < 1+\sqrt{2}.$$

Let $u(1+\sqrt{2})^{-k} = a + b\sqrt{2}$, so that

$$1 \leq a + b\sqrt{2} < 1+\sqrt{2}.$$

Since $a, b \in \mathbb{Z}$ it follows that $a = 1$ and $b = 0$. Thus we have that $u(1+\sqrt{2})^{-k} = 1$ and so $u = (1+\sqrt{2})^k$. Thus we see that the group of units coincides with the set $\{\pm(1+\sqrt{2})^k \mid k \geq 1\}$.

2.12 Let A be an ideal of R. Then the set

$$M = \{m \in \mathbb{Z} \mid \frac{m}{n} \in A \text{ for some } \frac{m}{n} \in R\}$$

forms an ideal of $\mathbb{Z}$. For, if $m_1, m_2 \in M$ then $\frac{m_1}{n_1}, \frac{m_2}{n_2} \in A$ whence $m_1, m_2 \in A$ since A is an ideal, and consequently $\frac{m_1 - m_2}{1} \in A$, giving $m_1 - m_2 \in M$. Also, if $a \in \mathbb{Z}$ then $\frac{a}{1}\frac{m_1}{n_1} \in A$ shows that $am_1 \in M$.

Now every ideal of $\mathbb{Z}$ is principal, so we have $M = (m)$, say. Thus if $m = p^k t$ where t is coprime to p then we must have $A = \left(\frac{p^k}{1}\right)$, showing that A is also principal. However, $\left(\frac{p}{1}\right)$ is a maximal ideal of R which contains A and so $\left(\frac{p}{1}\right)$ is the unique maximal ideal that contains every proper ideal of R.

2.13 (a) If x and $y_1 \cdots y_n$ are not relatively prime then there is an irreducible $d \in D$ such that $d|x$ and $d|y_1 \cdots y_n$. Since, in a principal ideal domain, irreducible and prime are the same we have $d|x$ and $d|y_i$ for some i. This contradicts the fact that x, y_i are given to be relatively prime.

(b) If $p_i^{r_i}$ and $p_j^{r_j}$ are not relatively prime then there is an irreducible (=prime) $p \in D$ such that $p|p_i^{r_i}$ and $p|p_j^{r_j}$ whence the contradiction $p_i \sim p \sim p_j$.

Let $b = p_1^{r_1} \cdots p_n^{r_n}$ be the unique factorisation of b as a product of irreducibles. By (b) the elements $p_1^{r_1}, \ldots, p_n^{r_n}$ are pairwise relatively prime. Hence, by (a), for every i the elements $p_i^{r_i}$ and $\prod_{j \neq i} p_j^{r_j}$ are relatively prime. It follows that the elements

$$\prod_{i \neq 1} p_i^{r_i}, \prod_{i \neq 2} p_i^{r_i}, \ldots, \prod_{i \neq n} p_i^{r_i}$$

are relatively prime. Thus there exist $\alpha_1, \ldots, \alpha_n$ such that

$$1 = \alpha_1 \prod_{i \neq 1} p_i^{r_i} + \cdots + \alpha_n \prod_{i \neq n} p_i^{r_i}.$$

Multiplying by a and dividing by $b = \prod_{i=1}^{n} p_i^{r_i}$ we obtain

$$x = \frac{a}{b} = \frac{a\alpha_1}{p_1^{r_1}} + \cdots + \frac{a\alpha_n}{p_n^{r_n}}.$$

2.14 (1) $\Rightarrow$ (2) : Let $\mathcal{F}$ be a family of ideals of A. Choose $I_0 \in \mathcal{F}$. Either I_0 is maximal in $\mathcal{F}$ or there exists $I_1 \in \mathcal{F}$ with $I_0 \subset I_1$. Either I_1 is maximal in $\mathcal{F}$ or there exists $I_2 \in \mathcal{F}$ with $I_0 \subset I_1 \subset I_2$. Since A is noetherian, this argument can be repeated only finitely many times, ending with a maximal ideal of $\mathcal{F}$.

(2) $\Rightarrow$ (3) : Let I be an ideal of A and let C be the collection of all ideals of A that are contained in I and are finitely generated. We have that $C \neq \emptyset$ since clearly $(0) \in C$. By (2), C has a maximal element J. For $x \in I$ consider the ideal $J + (x)$. This is finitely generated (since J is) and is an ideal of A contained in I, so belongs to C. The maximality of J then gives $J = J + (x)$ whence $x \in J$. Since x was chosen arbitrarily in I it follows that $I = J \in C$ and hence that I is finitely generated.

(3) $\Rightarrow$ (1) : Let $I_1 \subseteq I_2 \subseteq I_3 \subseteq \ldots$ be an increasing chain of ideals of A. By the hypothesis (3), every ideal of A is finitely generated. Let $I = \bigcup_{j \geq 1} I_j$. Then I is an ideal and so is finitely generated, say by

$\{x_1, \ldots, x_n\}$. Now every x_i belongs to some I_j, so let k be the greatest such j encountered. Then we have each $x_i \in I_k$ and it follows that

$$I = (x_1, \ldots, x_n) \subseteq I_k$$

whence $I = I_k$. Thus $I_k = I_{k+1} = I_{k+2} = \ldots$ and the chain terminates finitely.

$(\alpha) \Rightarrow (\beta)$: If A is a principal ideal domain then every ideal is principal so is finitely generated (in fact, generated by a singleton) so, from the implication (3) $\Rightarrow$ (1) above, A is noetherian. Also, the sum of two principal ideal is principal (since every ideal is principal).

$(\beta) \Rightarrow (\alpha)$: Suppose now that (β) holds and let I be an ideal of A. By (1) $\Rightarrow$ (3), I is finitely generated, say

$$I = (x_1, \ldots, x_n) = (x_1) + \cdots + (x_n).$$

A simple inductive argument shows that (by (β)) every finite sum of principal ideals is principal. Hence I is principal.

2.15 (a) $a|b$ implies that $b \in I = (a)$. But a is such that $N(a)$ is minimal for all elements of I. Thus, since $N(b) = N(a)$ and $b \in I$ we have $(a) = I = (b)$ and so a, b are associates.

(b) Let $a, b \in R$ be such that neither divides the other. Consider the ideal $I = (a, b)$ generated by a and b. Since $b \notin (a)$ and $a \notin (b)$ we see that $I \neq (a)$ and $I \neq (b)$. Now choose $d \in I$ with $N(d)$ minimal. Then $I = (d)$ and $N(d) < N(a), N(d) < N(b)$. From $(a, b) = (d)$ we deduce that $d \in (a, b)$ so there exist $\alpha, \beta \in R$ with $d = \alpha a + \beta b$ as required.

2.16 Suppose that there exist $a, b \in R \setminus \{0\}$ such that

$$\delta(a + b) > \max(\delta(a), \delta(b)).$$

Then from

$$b = 0(a + b) + b, \qquad b = 1(a + b) - a$$

with $\delta(-a) = \delta(a) < \delta(a + b)$ and $\delta(b) < \delta(a + b)$ we see the lack of uniqueness of quotient and remainder.

Conversely, suppose that $\delta(a + b) \leq \max(\delta(a), \delta(b))$ and that $a \in R$ has two representations, say

$$a = qb + r \qquad (r = 0 \text{ or } \delta(r) < \delta(b));$$
$$a = q'b + r' \qquad (r' = 0 \text{ or } \delta(r') < \delta(b)),$$

with $r \neq r'$ and $q \neq q'$. Then we have the contradiction

$$\delta(b) \leq \delta[(q-q')b] = \delta(r'-r) < \max(\delta(r'), \delta(-r)) < \delta(b).$$

Thus $r = r'$ or $q = q'$. Since each of these implies the other, uniqueness follows.

In $\mathbb{Z}[i]$ take $a = 1+4i$ and $b = 5+3i$. Then we have

$$\begin{aligned} 1+4i &= (5+3i)(1+i) + (-1-4i) \\ &= (5+3i)i + (4-i). \end{aligned}$$

2.17 (a) We have $-1+3i = (1+i)(1+2i)$. Since $\ell(1+i) = 2$ and $\ell(1+2i) = 5$ we see that $1+i$ and $1+2i$ are irreducible.

(b) Since g.c.d.$(a_0, \ldots, a_n) = 1$ we cannot write $f(X) = a\, f_1(X)$ with a irreducible. Hence if $f(X)$ is reducible we can suppose that $f(X) = g(X)h(X)$ where

$$\begin{aligned} g(X) &= b_0 + b_1 X + \cdots + b_r X^r, \\ h(X) &= c_0 + c_1 X + \cdots + c_s X^s, \end{aligned}$$

with $r, s < n$. Now $a_0 = b_0 c_0$ and so, since $p|a_0$, we have $p|b_0$ or $p|c_0$. Also, since p^2 does not divide $b_0 c_0$ we can assume that $p|b_0$ and p does not divide c_0. If $p|b_i$ for all i then $p|a_n$ which is a contradiction. Hence there is a first coefficient, b_t say, such that p does not divide b_t. But

$$a_t = (b_0 c_t + \cdots + b_{t-1} c_1) + b_t c_0$$

and p divides the bracketed part since $p|b_i$ for $0 \leq i < t$. Hence p does not divide a_t since it does not divide $b_t c_0$. Since $t < n$, this is a contradiction.

(c) Since $-1+3i = (1+i)(1+2i)$ we take $p = 1+i$ and note that $p|(-1+3i)$ but p^2 does not divide $-1+3i$. Also, since $(1+i)(1-i) = 2$ we see that $p|6$ and $p|8i$ but p does not divide 1. Hence, by (b), the polynomial

$$X^3 + 8iX^2 - 6X - 1 + 3i$$

is irreducible in $\mathbb{Z}[i][X]$.

2.18 $\mathbb{Z}[\sqrt{3}]$ is a euclidean domain so every pair of non-zero elements has a greatest common divisor.

We have, as products of irreducibles,

$$\begin{aligned} 13 &= (4-\sqrt{3})(4+\sqrt{3}), \\ 7+5\sqrt{3} &= (1+\sqrt{3})(4+\sqrt{3}). \end{aligned}$$

Hence the greatest common divisor is $4+\sqrt{3}$.

2.19 (a) $a^2+5b^2=\pm1$ if and only if $a=\pm1$ and $b=0$. So if $(a+b\sqrt{-5})(c+d\sqrt{-5})=1$ then

$$(a^2+5b^2)(c^2+5d^2)=\ell(a+b\sqrt{-5})\ell(c+d\sqrt{-5})=\ell(1)=1$$

gives $a^2+5b^2=1=c^2+5d^2$ and the result follows.

(b) $\ell(3)=\ell(2+\sqrt{-5})=\ell(2-\sqrt{-5})=9$, so if z is a factor of $3, 2+\sqrt{-5}, 2-\sqrt{-5}$ then we must have $\ell(z)\in\{1,3,9\}$. Now if $\ell(z)=3$ we have $z=a+b\sqrt{-5}$ where $a,b\in\mathbb{Z}$ are such that $a^2+5b^2=3$. Clearly, there are no solutions. Thus if z is a factor then we must have either $\ell(z)=1$ or $\ell(z)=9$. In either case, one of the two factors is of length 1 showing that it is a unit.

(c) We have that

$$9=3\cdot3=(2+\sqrt{-5})(2-\sqrt{-5}).$$

To show that these are different factorisations it suffices to observe that $3z\neq2+\sqrt{-5}$ for any unit z, for ±1 are the only units.

(d) To show that $P_1=(3,2+\sqrt{-5})$ is prime, consider $P_1\cap\mathbb{Z}$. We have $3\in P_1\cap\mathbb{Z}$, and if any integer not divisible by 3 lies in $P_1\cap\mathbb{Z}$ then, by the euclidean algorithm, $1\in P_1\cap\mathbb{Z}$ and we have the contradiction $P_1=\mathbb{Z}[\sqrt{-5}]$. Hence we see that $P_1\cap\mathbb{Z}=3\mathbb{Z}$.

Given $x,y\in\mathbb{Z}[\sqrt{-5}]$ let u,v be integers such that

$$x-u\in P_1,\quad y-v\in P_1.$$

Suppose that $xy\in P_1$. Then $uv\in P_1$ and, since $uv\in\mathbb{Z}$, we have $uv\in P_1\cap\mathbb{Z}=3\mathbb{Z}$. Thus $u\in3\mathbb{Z}$ or $v\in3\mathbb{Z}$, so either $x\in P_1$ or $y\in P_1$. Hence P_1 is prime. Similarly, so is $P_2=(3,2-\sqrt{-5})$.

(e) With the above notation, we have

$$\begin{aligned}P_1P_2&=(3,2+\sqrt{-5})(3,2-\sqrt{-5})\\&=(3)(3,2+\sqrt{-5},2-\sqrt{-5})\\&=(3),\end{aligned}$$

since $1=-3+(2+\sqrt{-5})+(2-\sqrt{-5})\in(3,2+\sqrt{-5},2-\sqrt{-5})$.

2.20 First note that

$$\begin{aligned}m+n\sqrt{10}\in P_1&\iff 2|m,\\m+n\sqrt{10}\in P_2&\iff m-n\equiv0\bmod 3,\\m+n\sqrt{10}\in P_3&\iff m+n\equiv0\bmod 3.\end{aligned}$$

Using this we can show that each of P_1, P_2, P_3 is prime. For example, if

$$(m_1 + n_1\sqrt{10})(m_2 + n_2\sqrt{10}) \in P_2$$

then we have

$$3 \mid [(m_1 - n_1)(m_2 - n_2) + 9n_1n_2]$$

and so $3|(m_1 - n_1)$ or $3|(m_2 - n_2)$.

Now $(6) = (2)(3) = (4 + \sqrt{10})(4 - \sqrt{10})$ and

$$(2) = P_1^2, \quad (3) = P_2P_3, \quad (4 + \sqrt{10}) = P_1P_2, \quad (4 - \sqrt{10}) = P_1P_3.$$

Hence both decompositions of the element 6 lead to the decomposition

$$(6) = P_1^2P_2P_3$$

for the ideal (6) as a product of prime ideals.

Solutions to Chapter 3

3.1 The binomial theorem (which holds in any commutative ring with a 1) gives

$$(a+b)^p = \sum_{r=0}^{p} \binom{p}{r} a^{p-r} b^r, \qquad (a-b)^p = \sum_{r=0}^{p} \binom{p}{r} (-1)^r a^{p-r} b^r.$$

For $0 < r < p$ the coefficient $\binom{p}{r} = \dfrac{p!}{r!(p-r)!}$ is an integer. Since $p|p!$ and p does not divide either $r!$ or $(p-r)!$, we see that $\binom{p}{r}$ is an integer that is divisible by p. Hence we have that

$$(a+b)^p = a^p + b^p$$

and, if p is odd,

$$(a-b)^p = a^p + (-1)^p b^p = a^p - b^p.$$

If p is even then we must have $p = 2$, in which case

$$(a-b)^2 = a^2 + b^2 = a^2 - b^2 + 2b^2 = a^2 - b^2.$$

For the second part, proceed by induction. The result has just been established for $n = 1$. For the inductive step, suppose that

$$(a \pm b)^{p^k} = a^{p^k} \pm b^{p^k}$$

where $k > 1$. Then we have

$$\begin{aligned}(a \pm b)^{p^{k+1}} = [(a \pm b)^{p^k}]^p &= (a^{p^k} \pm b^{p^k})^p \\ &= (a^{p^k})^p \pm (b^{p^k})^p \\ &= a^{p^{k+1}} \pm b^{p^{k+1}}.\end{aligned}$$

3.2 If F is of characteristic 0 then clearly

$$Df = 0 \iff a_1 = a_2 = \cdots = a_n = 0 \iff f = a_0.$$

If now F is of characteristic p then

(1) if $p|k$ we have

$$ka_k = (k1)a_k = 0a_k = 0;$$

(2) if p does not divide k we have

$$ka_k = 0 \Rightarrow (k1)a_k = 0 \Rightarrow a_k = 0.$$

Thus $Df = 0$ if and only if f is of the form

$$a_0 + a_pX^p + a_{2p}X^{2p} + \cdots + a_{rp}X^{rp}.$$

3.3 Clearly, every non-zero constant polynomial is a unit in $F[X]$. That these are all the units follows from the fact that if f, g are non-constant polynomials then $\deg fg \geq \deg f \geq 1$ so that fg is non-constant. Hence no polynomial of degree greater than zero is a unit.

If $\varphi : F[X] \to F[X]$ is an automorphism then φ sends units to units (since the units are the invertible elements). Hence the image under φ of a non-zero constant polynomial is also a non-zero constant polynomial.

Define $\vartheta : F \to F$ as follows : let $\vartheta(0) = 0$ and for every $a \neq 0$ let $\vartheta(a)$ be the constant polynomial $\varphi(a)$. Then clearly ϑ is an automorphism on F and the diagram

$$\begin{array}{ccc} F & \xrightarrow{i} & F[X] \\ \vartheta \downarrow & & \downarrow \varphi \\ F & \xrightarrow[i]{} & F[X] \end{array}$$

is commutative.

For the monomial $X \in F[X]$, consider $\varphi(X) \in F[X]$. If $\deg \varphi(X) < 1$ then $\varphi(X)$ is a constant polynomial so

$$\varphi(a_0 + a_1X + \cdots + a_nX^n) = \vartheta(a_0) + \vartheta(a_1)\varphi(X) + \cdots + \vartheta(a_n)(\varphi(X))^n$$

is also a constant polynomial, which is impossible since φ is an automorphism. If now $\deg \varphi(X) = k > 1$ then the image $\varphi(f)$ of a polynomial of degree n will have degree nk. Thus only polynomials of degrees $k, 2k, 3k, \ldots$ can appear as images under φ. Again this is impossible since φ is an automorphism. We must therefore have $\deg \varphi(X) = 1$ so that $\varphi(X) = aX + b$ for some $a \neq 0$.

3.4 We have that f, g are equivalent if and only if

$$(\forall \alpha \in F) \qquad (f - g)(\alpha) = 0,$$

which is the case if and only if $X - \alpha$ divides $f - g$ for all $\alpha \in F$. Suppose that $\deg(f - g) = n$.

If F is infinite and f, g are equivalent then taking $n + 1$ distinct elements $\alpha_0, \alpha_1, \ldots, \alpha_n$ in F we have that $f - g$ is divisible by

$$(X - \alpha_0)(X - \alpha_1) \cdots (X - \alpha_n)$$

which is of degree $n + 1$. Since a non-zero polynomial of degree n has at most n distinct roots, we conclude that $f - g = 0$ and so $f = g$.

If F is finite, say $F = \{a_1, a_2, \ldots, a_n\}$, then the above argument shows that $f - g$ is divisible by

$$m(X) = (X - a_1)(X - a_2) \cdots (X - a_n).$$

Conversely, if $f - g$ is divisible by $m(X)$ then it is clear that f and g are equivalent.

In the case where $F = \mathrm{GF}(p) = \mathbb{Z}/p\mathbb{Z}$ we have

$$m(X) = X(X - \bar{1})(X - \bar{2}) \cdots (X - \overline{p-1}).$$

Since, for $\alpha = \bar{0}, \bar{1}, \ldots, \overline{p-1}$ we have $\alpha^p = \alpha$ [recall Fermat's theorem that $a^{p-1} \equiv 1 \bmod p$], we have that $X - \alpha$ divides $X^p - X$. Hence $m(X)$ divides $X^p - X$, say $X^p - X = m(X)h(X)$. Equating degrees gives $p = p + \deg h$ and so $h(X)$ is a non-zero constant polynomial. Equating leading coefficients now gives $h(X) = 1$ and so $m(X) = X^p - X$.

3.5 Consider $X^2 + Y^2 - 1$ as an element of $F[Y][X] = F[X, Y]$, i.e. as a polynomial in X with coefficients in $F[Y]$.

Now $Y + 1$ is irreducible in $F[Y]$, and $Y + 1$ divides $Y^2 - 1$. Also, $(Y + 1)^2$ does not divide $Y^2 - 1$; for otherwise, since each is monic, we would have $(Y + 1)^2 = Y^2 - 1$ whence $2Y + 2.1 = 0$ which is possible only if F is of characteristic 2. We can therefore apply the result of question 2.17(b) to see that $X^2 + Y^2 - 1$ is irreducible in $F[X, Y]$.

3.6 The natural morphism $\mathbb{Z} \to \mathbb{Z}_n$ extends to a morphism $\mathbb{Z}[X] \to \mathbb{Z}_n[X]$; simply reduce the coefficients modulo n.

Given $f \in \mathbb{Z}[X]$, choose n such that n does not divide the leading coefficient of f. Then if f factorises over $\mathbb{Z}$ the image of f under the morphism factorises over $\mathbb{Z}_n$. Thus if the image of f is irreducible over $\mathbb{Z}_n$ then f must be irreducible over $\mathbb{Z}$.

The whole point behind this useful test for irreducibility is that $\mathbb{Z}_n$ is finite and so there are only finitely many possibilities to check.

In $\mathbb{Z}_5[X]$ we have $X^4 + 15X^3 + 7 = X^4 + 2$. This has no linear factor since no $t \in \mathbb{Z}_5$ satisfies $t^4 + 2 = 0$. Suppose that

$$X^4 + 2 = (X^2 + aX + b)(X^2 + cX + d).$$

Then we have

$$a + c = 0, \quad ac + b + d = 0, \quad ad + bc = 0, \quad bd = 2.$$

From $a = -c$ we have $a(b-d) = 0$. If $a = 0$ then $b = -d$ so $b^2 = -2 = 3$, while if $b = d$ then $b^2 = 2$. However, only $0, 1, 4$ are squares in $\mathbb{Z}_5$, so neither $b^2 = 3$ nor $b^2 = 2$ is possible. Hence we see that $X^4 + 2$ is irreducible over $\mathbb{Z}_5$. By the above, it is therefore irreducible over $\mathbb{Z}$. But if it were reducible over $\mathbb{Q}$ then it would be so over $\mathbb{Z}$.

3.7 $\{1, \sqrt{2}\}$ is a basis for $\mathbb{Q}(\sqrt{2})$ over $\mathbb{Q}$, and $\{1, \sqrt[3]{2}, (\sqrt[3]{2})^2\}$ is a basis for $\mathbb{Q}(\sqrt{2})(\sqrt[3]{2})$ over $\mathbb{Q}(\sqrt{2})$. Thus

$$\{1, \sqrt{2}, \sqrt[3]{2}, 2^{5/6}, 2^{2/3}, 2^{7/6}\}$$

is a basis for $\mathbb{Q}(\sqrt{2}, \sqrt[3]{2})$ over $\mathbb{Q}$.

Now $2^{7/6} = 2 \cdot 2^{1/6}$. Hence $2^{1/6} \in \mathbb{Q}(\sqrt{2}, \sqrt[3]{2})$ and so

$$\mathbb{Q} \subseteq \mathbb{Q}(\sqrt[6]{2}) \subseteq \mathbb{Q}(\sqrt{2}, \sqrt[3]{2}).$$

Now $\sqrt[6]{2}$ is a zero of $X^6 - 2$ which by Eisenstein's criterion is irreducible over $\mathbb{Q}$. This is therefore the minimum polynomial of $\sqrt[6]{2}$ over $\mathbb{Q}$. Hence we have

$$\begin{aligned} 6 = (\mathbb{Q}(\sqrt{2}, \sqrt[3]{2}) : \mathbb{Q}) &= \big(\mathbb{Q}(\sqrt{2}, \sqrt[3]{2}) : \mathbb{Q}(\sqrt[6]{2})\big)\big(\mathbb{Q}(\sqrt[6]{2}) : \mathbb{Q}\big) \\ &= \big(\mathbb{Q}(\sqrt{2}, \sqrt[3]{2}) : \mathbb{Q}(\sqrt[6]{2})\big) \cdot 6 \end{aligned}$$

so that $\big(\mathbb{Q}(\sqrt{2}, \sqrt[3]{2}) : \mathbb{Q}(\sqrt[6]{2})\big) = 1$ and hence $\mathbb{Q}(\sqrt{2}, \sqrt[3]{2}) = \mathbb{Q}(\sqrt[6]{2})$, a simple extension.

3.8 Suppose that Z were algebraic over F. Then there would exist a polynomial

$$f(X) = a_0 + a_1X + \cdots + a_nX^n \qquad (a_n \neq 0)$$

over F for which $f(Z) = 0$, i.e.

$$a_0 + a_1\left(\frac{X^3}{X+1}\right) + a_2\left(\frac{X^3}{X+1}\right)^2 + \cdots + a_n\left(\frac{X^3}{X+1}\right)^n = 0,$$

from which we have

$$a_nX^{3n} + a_{n-1}X^{3n-3}(X+1) + \cdots + a_0(X+1)^n = 0.$$

Since the left hand side is a non-zero polynomial of degree $3n$, we have a contradiction. Thus Z must be transcendental over F.

Now in $(F(Z))[Y]$ we have that X satisfies $Y^3 - ZY - Z = 0$. But $Y^3 - ZY - Z$ is irreducible in $(F(Z))[Y]$; for, if not, it has a linear factor $Y - \alpha$ where $\alpha \in F(Z)$ and $\alpha^3 - Z\alpha - Z = 0$, and this would imply the existence of a non-zero polynomial f for which $f(Z) = 0$ which, as we have just seen, is impossible. Thus we see that $Y^3 - ZY - Z$ is the mimimum polynomial of X over $F(Z)$. Thus $F(X)$ is a simple algebraic extension of $F(Z)$.

3.9 Both are irreducible by the Eisenstein criterion. We have

$$X^2 - 3 = (X - \sqrt{3})(X + \sqrt{3})$$
$$X^2 - 2X - 2 = (X - 1 - \sqrt{3})(X - 1 + \sqrt{3})$$

and the splitting field in each case is $\mathbb{Q}(\sqrt{3})$.

3.10 Over $\mathbb{C}$ we have

$$(X^2 - 2X - 2)(X^2 + 1) = (X - 1 + \sqrt{3})(X - 1 - \sqrt{3})(X - i)(X + i).$$

Hence a splitting field is $\mathbb{Q}(\sqrt{3}, i)$.

We also have

$$\begin{aligned} &X^5 - 3X^3 + X^2 - 3 \\ &= (X^2 - 3)(X^3 + 1) \\ &= (X - \sqrt{3})(X + \sqrt{3})(X + 1)(X - \tfrac{1}{2}(-1 + i\sqrt{3}))(X - \tfrac{1}{2}(-1 - i\sqrt{3})) \end{aligned}$$

and a splitting field is

$$\mathbb{Q}(\sqrt{3}, \tfrac{1}{2}(-1 + i\sqrt{3})) = \mathbb{Q}(\sqrt{3}, i).$$

Let $K = \mathbb{Q}(\sqrt{3}, i)$. Now $\sqrt{3}$ has minimum polynomial $X^2 - 3$ over $\mathbb{Q}$ so $(\mathbb{Q}(\sqrt{3}) : \mathbb{Q}) = 2$. Since $i \notin \mathbb{Q}(\sqrt{3})$ and since i has minimum polynomial $X^2 + 1$ over $\mathbb{Q}(\sqrt{3})$ we have $(\mathbb{Q}(\sqrt{3}, i) : \mathbb{Q}(\sqrt{3})) = 2$. Hence

$$(K : \mathbb{Q}) = (K : \mathbb{Q}(\sqrt{3}))(\mathbb{Q}(\sqrt{3}) : \mathbb{Q}) = 2 \cdot 2 = 4.$$

3.11 It is readily seen that

$$X^3 - 3abX + a^3 + b^3 = (X + a + b)(X^2 + (a + b)X + (a + b)^2 - 3ab).$$

Since $X^6 - 6X^3 + 8 = (X^3 - 2)(X^3 - 4)$, a splitting field over $\mathbb{Q}$ is $\mathbb{Q}(\sqrt[3]{2}, \sqrt[3]{4}, i\sqrt{3})$. But $(\sqrt[3]{2})^2 = \sqrt[3]{4}$ so $\mathbb{Q}(\sqrt[3]{2}, \sqrt[3]{4}, i\sqrt{3}) = \mathbb{Q}(\sqrt[3]{2}, i\sqrt{3})$. Hence $S = \mathbb{Q}(\sqrt[3]{2}, i\sqrt{3})$ and

$$(S : \mathbb{Q}) = (\mathbb{Q}(\sqrt[3]{2}, i\sqrt{3}) : \mathbb{Q}(\sqrt[3]{2}))(\mathbb{Q}(\sqrt[3]{2}) : \mathbb{Q}) = 2 \cdot 3 = 6.$$

Since S contains $\sqrt[3]{2}$ and $\sqrt[3]{4}$ we have $\sqrt[3]{2} + \sqrt[3]{4} \in S$. Now take $a = -\sqrt[3]{4}, b = -\sqrt[3]{2}$ in the first part of the question to see that

$$X - (\sqrt[3]{4} + \sqrt[3]{2}) \text{ divides } X^3 - 6X - 6.$$

Since $\sqrt[3]{4} + \sqrt[3]{2} \in S$, which is of degree 3 over $\mathbb{Q}$, and since $\sqrt[3]{4} + \sqrt[3]{2} \notin \mathbb{Q}$, we see that the minimum polynomial of $\sqrt[3]{4} + \sqrt[3]{2}$ over $\mathbb{Q}$ is a cubic. By the above observation, it must therefore be $X^3 - 6X - 6$.

Since $\sqrt[3]{4} + \sqrt[3]{2} \in S$ we have $\mathbb{Q}(\sqrt[3]{4} + \sqrt[3]{2}) \subseteq S$. But $\mathbb{Q}(\sqrt[3]{4} + \sqrt[3]{2})$ has degree 3 over $\mathbb{Q}$ while S has degree 6, so we conclude that $\mathbb{Q}(\sqrt[3]{4} + \sqrt[3]{2}) \neq S$.

3.12 In $\mathbb{Q}[X]$ we have

$$f(X) = (X^2 + 4X + 1)(X^2 - 2X - 1)$$

and in $\mathbb{R}[X]$ we have

$$f(X) = (X + 2 - \sqrt{3})(X + 2 + \sqrt{3})(X - 1 - \sqrt{2})(X - 1 + \sqrt{2}).$$

Thus $K = \mathbb{Q}(\sqrt{2}, \sqrt{3})$ is a splitting field for f.

Now if we had $\sqrt{3} \in \mathbb{Q}(\sqrt{2})$ then there would exist $a, b \in \mathbb{Q}$ with $\sqrt{3} = a + b\sqrt{2}$ whence

$$3 = a^2 + 2b^2 + 2ab\sqrt{2}.$$

If $a, b \neq 0$ we have the contradiction $\sqrt{2} \in \mathbb{Q}$. If $a = 0$ then $\sqrt{3} = b\sqrt{2}$ so $\sqrt{6} = 2b \in \mathbb{Q}$, another contradiction. If $b = 0$ then $\sqrt{3} = a \in \mathbb{Q}$, also a contradiction. Thus we see that $\sqrt{3} \notin \mathbb{Q}(\sqrt{2})$ and so

$$(\mathbb{Q}(\sqrt{2},\sqrt{3}) : \mathbb{Q}) = (\mathbb{Q}(\sqrt{2},\sqrt{3}) : \mathbb{Q}(\sqrt{2}))(\mathbb{Q}(\sqrt{2}) : \mathbb{Q}) = 2 \cdot 2 = 4.$$

A basis for $\mathbb{Q}(\sqrt{2},\sqrt{3})$ over $\mathbb{Q}$ is

$$\{1, \sqrt{2}, \sqrt{3}, \sqrt{6}\}.$$

Now the minimum polynomial of $\sqrt{3} - \sqrt{2}$ has degree 2 or 4. If it were of degree 2 then for some $a, b \in \mathbb{Q}$ we would have

$$(\sqrt{3} - \sqrt{2})^2 + a(\sqrt{3} - \sqrt{2}) + b = 0$$

which gives

$$b + 5 + a\sqrt{3} - a\sqrt{2} - 2\sqrt{6} = 0,$$

which contradicts the fact that $\{1, \sqrt{2}, \sqrt{3}, \sqrt{6}\}$ is a basis. Hence we see that the minimum polynomial has degree 4 and $\mathbb{Q}(\sqrt{2},\sqrt{3}) = \mathbb{Q}(\sqrt{3} - \sqrt{2})$.

The minimum polynomial of $\sqrt{3}-\sqrt{2}$ over $\mathbb{Q}$ is $X^4 - 10X^2 + 1$ whereas over $\mathbb{Q}(\sqrt{3})$ it is $X^2 - 2\sqrt{3}X + 1$.

3.13 We have that

$$X^4 - X^2 - 2 = (X^2 - 2)(X^2 + 1)$$

with $X^2 - 2$ and $X^2 + 1$ irreducible over $\mathbb{Q}$. Now over $\mathbb{C}$

$$X^4 - X^2 - 2 = (X - \sqrt{2})(X + \sqrt{2})(X - i)(X + i)$$

so f splits completely over $\mathbb{Q}(\sqrt{2}, i)$. Since clearly f cannot split in any smaller subfield of $\mathbb{C}$, we have that $K = \mathbb{Q}(\sqrt{2}, i)$ is a splitting field.

Now $\sqrt{2}$ has minimum polynomial $X^2 - 2$ over $\mathbb{Q}$ and so we have $\mathbb{Q}(\sqrt{2}) : \mathbb{Q}) = 2$. Since $i \notin \mathbb{Q}(\sqrt{2})$ and since i has minimum polynomial $X^2 + 1$ over $\mathbb{Q}(\sqrt{2})$ we also have that $(\mathbb{Q}(\sqrt{2}, i) : \mathbb{Q}(\sqrt{2})) = 2$. Hence

$$(K : \mathbb{Q}) = (K : \mathbb{Q}(\sqrt{2}))(\mathbb{Q}(\sqrt{2}) : \mathbb{Q}) = 2 \cdot 2 = 4.$$

A basis for K over $\mathbb{Q}$ is $\{1, \sqrt{2}, i, i\sqrt{2}\}$.

Since $\mathbb{Q}(i+\sqrt{2}) \subseteq \mathbb{Q}(i,\sqrt{2})$ we have that $(\mathbb{Q}(i+\sqrt{2}):\mathbb{Q})$ is either 2 or 4. Now $(i+\sqrt{2})^2 = 1+2i\sqrt{2}$, and if the minimum polynomial of $i+\sqrt{2}$ were of degree 2 we could find $a,b \in \mathbb{Q}$ such that

$$\begin{aligned} 0 &= (i+\sqrt{2})^2 + a(i+\sqrt{2}) + b \\ &= b + 1 + a\sqrt{2} + ai + 2i\sqrt{2}, \end{aligned}$$

which contradicts the fact that $\{1,\sqrt{2},i,i\sqrt{2}\}$ is a basis. Thus the minimum polynomial of $i+\sqrt{2}$ is of degree 4 and so $(\mathbb{Q}(i+\sqrt{2}):\mathbb{Q}) = 4$ and consequently $\mathbb{Q}(i+\sqrt{2}) = \mathbb{Q}(i,\sqrt{2})$.

Since clearly $(i+\sqrt{2})^2 = 1+2i\sqrt{2}$ we have $(i+\sqrt{2})^4 = (1+2i\sqrt{2})^2 = -7+4i\sqrt{2}$ and so

$$(i+\sqrt{2})^4 - 2(i+\sqrt{2})^2 + 9 = 0,$$

whence the minimum polynomial of $i+\sqrt{2}$ is $X^4 - 2X^2 + 9$.

(a) Since $(i+\sqrt{2})^2 = 1+2i\sqrt{2} = 1+2i(i+\sqrt{2})+2$ the minimum polynomial of $i+\sqrt{2}$ over $\mathbb{Q}(i)$ is $X^2 - 2iX - 3$.

(b) Since $(i+\sqrt{2})^2 = 1+2i\sqrt{2} = 1+2\sqrt{2}(i+\sqrt{2}) - 4$ the minimum polynomial of $i+\sqrt{2}$ over $\mathbb{Q}(\sqrt{2})$ is $X^2 - 2\sqrt{2}X + 3$.

(c) Since $(i+\sqrt{2})^2 = 1+2i\sqrt{2}$ the minimum polynomial of $i+\sqrt{2}$ over $\mathbb{Q}(i\sqrt{2})$ is $X^2 - 1 - 2i\sqrt{2}$.

3.14 (1) $\Rightarrow$ (2) : If f is reducible then it has a linear or a quadratic factor. If it has a linear factor $X-\alpha$ then α is a root, whence so is $-\alpha$, whence f has the factor $X+\alpha$. Thus f has the quadratic factor $(X+\alpha)(X-\alpha)$.

(2) $\Rightarrow$ (3) : If $X^4 + r = (X^2+\alpha X+\beta)(X^2+\alpha' X+\beta')$ then

$$\alpha+\alpha' = 0,\ \beta+\beta'+\alpha\alpha' = 0,\ \alpha\beta'+\alpha'\beta = 0,\ \beta\beta' = r.$$

If $\alpha \neq 0$ then $\alpha' = -\alpha$ so $\alpha(\beta'-\beta) = 0$ so $\beta' = \beta$, which gives $r = \beta^2$ with $2\beta = -\alpha\alpha' = \alpha^2$, i.e. $r = \frac{1}{4}\alpha^4$.

If $\alpha = 0$ then $\alpha' = 0, \beta' = -\beta$ and $r = -\beta^2$.

(3) $\Rightarrow$ (1) : If $r = -p^2$ then $X^4 + r = X^4 - p^2 = (X^2+p)(X^2-p)$; and if $r = \frac{1}{4}q^4$ then

$$X^4 + r = X^4 + \tfrac{1}{4}q^4 = (X^2+qX+\tfrac{1}{2}q^2)(X^2-qX+\tfrac{1}{2}q^2).$$

Every element of $K = \mathbb{Q}(\xi)$ is of the form

$$\eta = a + b\xi + c\xi^2 + d\xi^3$$

where $a, b, c, d \in \mathbb{Q}$ and $\xi^4 = -r \in \mathbb{Q}$. Now $(\mathbb{Q}(\xi) : \mathbb{Q}) = 4$ so if $\mathbb{Q} \subset L \subset \mathbb{Q}(\xi)$ we must have $(L : \mathbb{Q}) = 2$. Thus L is necessarily of the form $\mathbb{Q}(\eta)$ where $\eta^2 \in \mathbb{Q}$ and $\eta \notin \mathbb{Q}$. Now

$$\begin{aligned}\eta^2 &= a^2 + b^2\xi^2 + c^2\xi^4 + d^2\xi^6 + 2ab\xi + 2ac\xi^2 \\ &\qquad + 2ad\xi^3 + 2bc\xi^3 + 2bd\xi^4 + 2cd\xi^5 \\ &= a^2 + b^2\xi^2 - rc^2 - rd^2\xi^2 + 2ab\xi + 2ac\xi^2 \\ &\qquad + 2ad\xi^3 + 2bc\xi^3 - 2bdr - 2cdr\xi \\ &= (a^2 - rc^2 - 2rbd) + 2(ab - rcd)\xi \\ &\qquad + (b^2 - rd^2 + 2ac)\xi^2 + 2(ad + bc)\xi^3,\end{aligned}$$

so $\eta^2 \in \mathbb{Q}, \eta \notin \mathbb{Q}$ is equivalent to

$$\begin{aligned}&ad + bc = 0,\\ &ab - rcd = 0,\\ &b^2 - rd^2 + 2ac = 0,\\ &b, c, d \text{ not all } 0.\end{aligned}$$

Now if $c \neq 0$ then since $r \neq -b^2/d^2$ we must have $a = b = d = 0$ in which case $\eta = c\xi^2$ and $\mathbb{Q}(\eta) = \mathbb{Q}(c\xi^2) = \mathbb{Q}(\xi^2)$. On the other hand, if $c = 0$ then

$$\begin{aligned}&ad = 0,\\ &ab = 0,\\ &b^2 = rd^2,\\ &b, d \text{ not both } 0.\end{aligned}$$

Thus if $\sqrt{r} \notin \mathbb{Q}$ there are no further solutions. If, however, $\sqrt{r} \in \mathbb{Q}$ then these conditions give

$$a = 0, \quad \frac{b}{d} = \pm r.$$

In this case we can take $d = 1$ and $b = \pm\sqrt{r}$ so that there are two other solutions, namely

$$\eta = \sqrt{r}\xi + \xi^3 \quad \text{and} \quad \eta = -\sqrt{r}\xi + \xi^3.$$

3.15 (a) The Galois group has order 4 since the minimum polynomial of $\omega = e^{2\pi i/5}$ is $X^4 + X^3 + X^2 + X + 1$.

(b) The order of the Galois group is 1 (i.e. every $\mathbb{Q}$–automorphism of $\mathbb{Q}(\omega)$ is the identity). For, suppose that α is a $\mathbb{Q}$–automorphism. Then since $\omega = \sqrt[3]{2}$ we have

$$[\alpha(\omega)]^3 = \alpha(\omega^3) = \alpha(2) = 2.$$

But since $\mathbb{Q}(\omega) \subseteq \mathbb{R}$ we have $\alpha(\omega) \in \mathbb{R}$ and so $\alpha(\omega) = \omega$. Hence α is the identity map.

3.16 In $\mathbb{C}[X]$ we have

$$X^4 - 2 = (X + r)(X - r)(X + ir)(X - ir)$$

where $r = \sqrt[4]{2}$. Hence a splitting field for $X^4 - 2$ is $\mathbb{Q}(\sqrt[4]{2}, i)$. Also,

$$(\mathbb{Q}(\sqrt[4]{2}, i) : \mathbb{Q}(\sqrt[4]{2}))(\mathbb{Q}(\sqrt[4]{2}) : \mathbb{Q}) = 2 \cdot 4 = 8.$$

A $\mathbb{Q}$–automorphism of $\mathbb{Q}(\sqrt[4]{2}, i)$ is completely determined by its effect on $\sqrt[4]{2}$ and i. Now the conjugates of i are $i, -i$ since its minimum polynomial is $X^2 + 1$. Also, the conjugates of $\sqrt[4]{2}$ are

$$\sqrt[4]{2},\ -\sqrt[4]{2},\ i\sqrt[4]{2},\ -i\sqrt[4]{2}$$

since its minimum polynomial is $X^4 - 2$. Hence there are eight elements in the Galois group

$$\mathrm{Gal}(\mathbb{Q}(\sqrt[4]{2}, i), \mathbb{Q}).$$

The elements α, β given by $\alpha(r) = -r, \alpha(i) = -i$ and $\beta(r) = ir, \beta(i) = i$ do not commute, so the Galois group is not abelian.

3.17 Let $\alpha = \sqrt[3]{2}$. Then

$$\begin{aligned} X^3 - 2 &= (X - \alpha)(X^2 + \alpha X + \alpha^2) \\ &= (X - \alpha)\big(X - \alpha(-\tfrac{1}{2} + \tfrac{i\sqrt{3}}{2})\big)\big(X - \alpha(-\tfrac{1}{2} - \tfrac{i\sqrt{3}}{2})\big). \end{aligned}$$

By the uniqueness theorem, any splitting field of $X^2 - 2$ is isomorphic to the subfield of $\mathbb{C}$ generated over $\mathbb{Q}$ by

$$\{\alpha, (-\tfrac{1}{2} + \tfrac{i\sqrt{3}}{2})\alpha, (-\tfrac{1}{2} - \tfrac{i\sqrt{3}}{2})\alpha\}.$$

All these elements belong to $\mathbb{Q}(\alpha, i\sqrt{3})$; and since

$$i\sqrt{3} = \alpha^{-1}[(-\tfrac{1}{2} + \tfrac{i\sqrt{3}}{2})\alpha - (-\tfrac{1}{2} - \tfrac{i\sqrt{3}}{2})\alpha]$$

it follows that a splitting field is $\mathbb{Q}(\sqrt[3]{2}, i\sqrt{3})$.

Now $(\mathbb{Q}(\sqrt[3]{2}, i\sqrt{3}) : \mathbb{Q})$ is a multiple of $(\mathbb{Q}(\sqrt[3]{2}) : \mathbb{Q})$ and $(\mathbb{Q}(i\sqrt{3} : \mathbb{Q})$, so is a multiple of 6. But $X^3 - 2$ is of degree 3, so the splitting field has degree at most $3! = 6$. Hence $(\mathbb{Q}(\sqrt[3]{2}, i\sqrt{3}) : \mathbb{Q}) = 6$. Thus the Galois group of $X^3 - 2$ over $\mathbb{Q}$ is of order 6.

Now let $K = \mathbb{Q}(\sqrt[3]{2}, i\sqrt{3})$. A $\mathbb{Q}$-automorphism of K is completely determined by its action on $\sqrt[3]{2}$ and on $i\sqrt{3}$. The conjugates of $\alpha = \sqrt[3]{2}$ are

$$\alpha, (-\tfrac{1}{2} + \tfrac{i\sqrt{3}}{2})\alpha, (-\tfrac{1}{2} - \tfrac{i\sqrt{3}}{2})\alpha$$

and those of $i\sqrt{3}$ are $i\sqrt{3}, -i\sqrt{3}$. So the six elements of $\mathrm{Gal}(K, \mathbb{Q})$ are represented by

$$e = \begin{pmatrix} \alpha & i\sqrt{3} \\ \alpha & i\sqrt{3} \end{pmatrix}, \qquad a = \begin{pmatrix} \alpha & i\sqrt{3} \\ (-\frac{1}{2} + \frac{i\sqrt{3}}{2})\alpha & i\sqrt{3} \end{pmatrix},$$

$$b = \begin{pmatrix} \alpha & i\sqrt{3} \\ (-\frac{1}{2} - \frac{i\sqrt{3}}{2})\alpha & i\sqrt{3} \end{pmatrix}, \qquad g = \begin{pmatrix} \alpha & i\sqrt{3} \\ \alpha & -i\sqrt{3} \end{pmatrix},$$

$$h = \begin{pmatrix} \alpha & i\sqrt{3} \\ (-\frac{1}{2} + \frac{i\sqrt{3}}{2})\alpha & -i\sqrt{3} \end{pmatrix}, \qquad k = \begin{pmatrix} \alpha & i\sqrt{3} \\ (-\frac{1}{2} - \frac{i\sqrt{3}}{2})\alpha & -i\sqrt{3} \end{pmatrix}.$$

The group table is

$\circ$	e	a	b	g	h	k
e	e	a	b	g	h	k
a	a	b	e	h	k	g
b	b	e	a	k	g	h
g	g	k	h	e	b	a
h	h	g	k	a	e	b
k	k	h	g	b	a	e

The subgroups are

$$\{e\},\ \{e, a, b\},\ \{e, g\},\ \{e, h\},\ \{e, k\},$$

and the associated Hasse diagram is

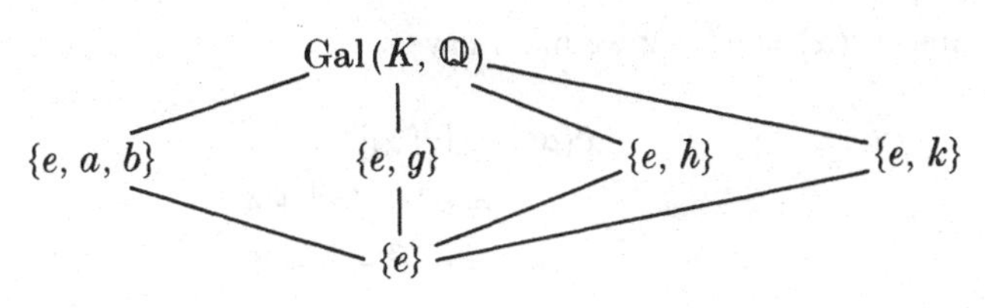

By the fundamental theorem of Galois theory, the subfields of the splitting field have a lattice diagram that is the dual of the above subgroup lattice diagram. As $K = \mathbb{Q}(\sqrt[3]{2}, i\sqrt{3})$ it must have four subfields (other than K and $\mathbb{Q}$). These are

$\mathbb{Q}(\sqrt[3]{2})$ of degree 3 —the fixed field of $\{e, g\}$;

$\mathbb{Q}(i\sqrt{3})$ of degree 2 —the fixed field of $\{e, a, b\}$;

$\mathbb{Q}((-\frac{1}{2} - \frac{i\sqrt{3}}{2})\alpha)$ of degree 3 —the fixed field of $\{e, h\}$;

$\mathbb{Q}((-\frac{1}{2} + \frac{i\sqrt{3}}{2})\alpha)$ of degree 3 —the fixed field of $\{e, k\}$.

3.18 First note that since α is a zero of f we have $\alpha^3 = 3\alpha - 1$. Now

$$\begin{aligned}
&(X-\alpha)(X-\alpha^2+2)(X+\alpha^2+\alpha-2)\\
&= (X^2+(-\alpha^2-\alpha+2)X+\alpha^3-2\alpha)(X+\alpha^2+\alpha-2)\\
&= (X^2+(-\alpha^2-\alpha+2)X+\alpha-1)(X+\alpha^2+\alpha-2)\\
&= X^3+(-\alpha^2-\alpha+2+\alpha^2+\alpha-2)X^2\\
&\qquad +[(-\alpha^2-\alpha+2)(\alpha^2+\alpha-2)+\alpha-1]X+(\alpha-1)(\alpha^2+\alpha-2).
\end{aligned}$$

But the coefficient of X^2 is clearly 0, whereas that of X is

$$\begin{aligned}
&-\alpha^4-\alpha^3+2\alpha^2-\alpha^3-\alpha^2+2\alpha+2\alpha^2+2\alpha-4+\alpha-1\\
&= -\alpha(3\alpha-1)-2(3\alpha-1)+3\alpha^2+5\alpha-5\\
&= -3,
\end{aligned}$$

and the constant term is

$$\alpha^3+\alpha^2-2\alpha-\alpha^2-\alpha+2 = 1.$$

Since $f(X) = (X-\alpha)(X-\alpha^2+2)(X+\alpha^2+\alpha-2)$ we see that $\mathbb{Q}(\alpha)$ is a splitting field for f over $\mathbb{Q}$ and that $\{1, \alpha, \alpha^2\}$ is a basis. Hence we have $(\mathbb{Q}(\alpha) : \mathbb{Q}) = 3$.

Since $\vartheta(\alpha) = \alpha^2 - 2$ we must have

$$\begin{aligned}
\vartheta(\alpha^2) &= [\vartheta(\alpha)]^2\\
&= \alpha^4-4\alpha^2+4\\
&= -\alpha^2-\alpha+4
\end{aligned}$$

since $\alpha^3 = 3\alpha - 1$. Hence ϑ extends to a unique $\mathbb{Q}$–automorphism defined by

$$\begin{aligned}\vartheta(a + b\alpha + c\alpha^2) &= a + b(\alpha^2 - 2) + c(-\alpha^2 - \alpha + 4) \\ &= (a - 2b + 4c) + (-c)\alpha + (b - c)\alpha^2.\end{aligned}$$

Therefore $\mathrm{Gal}(\mathbb{Q}(\alpha), \mathbb{Q}) = \{1, \vartheta, \vartheta^2\}$ is the cyclic group of order 3.

Note that any $\mathbb{Q}$–automorphism that sends α to $\alpha^2 - 2$ sends $\alpha^2 - 2$ to $(-\alpha^2 - \alpha + 4) - 2 = -\alpha^2 - \alpha + 2$, so no $\mathbb{Q}$–automorphism interchanges α and $\alpha^2 - 2$.

3.19 Only (a) is normal, a splitting field of $X^2 - 2$. All of the others have minimum polynomials with complex roots.

3.20 (a) We have $f(X) = (X - X_1)(X - X_2)\cdots(X - X_n)$ and f does not split in any proper subfield of K.

(b) An F–automorphism of K maps conjugates to conjugates, so permutes the X_i. Conversely, if $\psi \in S_n$, the symmetric group of degree n, then the mapping $\psi^* : K \to K$ such that

$$\begin{aligned}\psi^*(a) &= a \quad (a \in E) \\ \psi^*(X_i) &= X_{\psi(i)} \quad (i = 1, \ldots, n)\end{aligned}$$

defines an F–automorphism on K since the symmetric polynomials are unaltered. Thus we see that $\psi \mapsto \psi^*$ describes an isomorphism from S_n onto $\mathrm{Gal}(K, F)$.

(c) $(K : F) = |\mathrm{Gal}(K, F)| = |S_n| = n!$.

Solutions to Chapter 4

4.1 This involves simply a trivial verification of the axioms; for example,

$$(f+g)m = (f+g)(m) = f(m) + g(m) = fm + gm.$$

4.2 $\Rightarrow$: Define $\mu_r : M \to M$ by $\mu_r(m) = rm$. Then $r \mapsto \mu_r$ is a 1–preserving ring morphism $\mu : R \to \operatorname{End} M$.

$\Leftarrow$: Define an external law by $(r, m) \mapsto rm = [\mu(r)](m)$. Then since μ is a ring morphism we have

$r(m_1 + m_2) = rm_1 + rm_2,$
$(r+s)m = rm + sm,$
$(rs)m = r(sm),$
$1m = m,$

and so M is an R–module.

4.3 Take $R = M = \mathbb{Z}$ in question 4.2. If $f \in \operatorname{End} \mathbb{Z}$ then f is completely determined by $f(1)$ since $f(m) = f(m1) = mf(1)$ for every $m \in \mathbb{Z}$. Now observe that $\mu_{f(1)} = f$ so that μ is surjective :

$$(\forall m \in \mathbb{Z}) \qquad \mu_{f(1)}(m) = f(1)m = f(m).$$

If now $r \in \operatorname{Ker} \mu$ then $\mu(r) = 0$ gives $rm = 0$ for all $m \in \mathbb{Z}$ whence $r = 0$. Thus μ is a ring isomorphism.

The result for $\mathbb{Q}$ is similar : here we use the fact that if $q = \dfrac{n}{m} \in \mathbb{Q}$ then $mq = n$ so $mf(q) = f(mq) = f(n) = nf(1)$ whence

$$f(q) = \frac{n}{m} f(1) = qf(1).$$

4.4 If $\alpha, \beta \in R$ exist such that $f(x, y) = \alpha x + \beta y$ for all $x, y \in R$ then clearly f is an R–morphism.

Conversely, if $f : R \times R \to R$ is an R-morphism then for all $x, y \in R$ we have

$$f(x,y) = f[x(1,0) + y(0,1)] = xf(1,0) + yf(0,1)$$

whence, since R is commutative, the result follows with $\alpha = f(1,0)$ and $\beta = f(0,1)$.

4.5 (a) Not a ring morphism; for example,

$$f(\sqrt{2} \cdot \sqrt{2}) = f(2) = 2 \quad \text{and} \quad f(\sqrt{2})f(\sqrt{2}) = 1 \cdot 1 = 1.$$

(b) Not a $\mathbb{Z}[\sqrt{2}]$-morphism; for example,

$$\sqrt{2}f(1) = \sqrt{2} \quad \text{and} \quad f(\sqrt{2} \cdot 1) = f(\sqrt{2}) = 1.$$

(c) f is, however, a $\mathbb{Z}$-morphism since

$$f[(a + b\sqrt{2}) + (c + d\sqrt{2})] = a + b + c + d = f(a + b\sqrt{2}) + f(c + d\sqrt{2}),$$

$$f[n(a + b\sqrt{2})] = na + nb = n(a + b) = nf(a + b\sqrt{2}).$$

4.6 That $\lambda f \in \mathrm{Mor}_R(R, M)$ and that $\mathrm{Mor}_R(R, M)$ is an R-module is routine. That $\vartheta : \mathrm{Mor}_R(R, M) \to M$ given by $\vartheta(f) = f(1_R)$ is an R-morphism follows from the fact that

$$\vartheta(f + g) = (f + g)(1_R) = f(1_R) + g(1_R) = \vartheta(f) + \vartheta(g),$$

$$\vartheta(\lambda f) = (\lambda f)(1_R) = f(1_R\lambda) = f(\lambda) = \lambda f(1_R) = \lambda\vartheta(f).$$

That ϑ is injective follows from the observation that if $\vartheta(f) = \vartheta(g)$ then $f(1_R) = g(1_R)$ and so f, g agree on the basis $\{1_R\}$ of the R-module R. Thus we have $f = g$. Finally, given $m \in M$ let $f_m : R \to M$ be defined by $f_m(r) = rm$. Clearly, we have $f_m \in \mathrm{Mor}_R(R, M)$ and $m = 1_R m = f_m(1_R) = \vartheta(f_m)$. Consequently ϑ is also surjective.

4.7 Observe first that $f^{\rightarrow}$ and $f^{\leftarrow}$ are inclusion-preserving. Since $\mathrm{Ker}\, f = f^{\leftarrow}\{0\}$ and $\mathrm{Im}\, f = f^{\rightarrow}(M)$ it is then clear that

$$A + \mathrm{Ker}\, f \subseteq f^{\leftarrow}[f^{\rightarrow}(A)],$$

$$f^{\rightarrow}[f^{\leftarrow}(B)] \subseteq B \cap \mathrm{Im}\, f.$$

(a) If now $x \in f^{\leftarrow}[f^{\rightarrow}(A)]$ then we have $f(x) = f(a)$ for some $a \in A$ whence $x - a \in \mathrm{Ker}\, f$ and $x \in A + \mathrm{Ker}\, f$. This establishes the first equality.

(b) If $x \in B \cap \mathrm{Im}\, f$ then for some $y \in M$ we have $f(y) = x \in B$. This gives $y \in f^{\leftarrow}(B)$ and so $x = f(y) \in f^{\rightarrow}[f^{\leftarrow}(B)]$, which establishes the second equality.

(c) Since $f^{\rightarrow}$ is inclusion-preserving we have $f^{\rightarrow}(A \cap f^{\leftarrow}(B)) \subseteq f^{\rightarrow}(A)$ and $f^{\rightarrow}(A \cap f^{\leftarrow}(B)) \subseteq f^{\rightarrow}[f^{\leftarrow}(B)] \subseteq B$. Thus it suffices to prove that if $x \in f^{\rightarrow}(A) \cap B$ then $x = f(a)$ for some $a \in A \cap f^{\leftarrow}(B)$; and this is immediately seen from $x = f(a) \in B$.

4.8 (a) Let $i : \mathbb{Z} \to \mathbb{Q}$ be the canonical inclusion and let M be a non-zero submodule of $\mathbb{Q}$. Then there is a non-zero rational $\frac{m}{n} \in M$. Thus $m = n \cdot \frac{m}{n} \in M$ and so $M \cap \mathbb{Z} \neq \{0\}$. Consequently $i^{\leftarrow}(M) \neq \{0\}$ and i is essential.

(b) Let $j : \mathbb{Q} \to \mathbb{R}$ be the canonical inclusion. Let $\alpha \in \mathbb{R}$ be irrational. Then $\alpha\mathbb{Z}$ is a non-zero submodule of $\mathbb{R}$. Clearly, $\alpha\mathbb{Z} \cap \mathbb{Q} = \{0\}$ so $j^{\leftarrow}(\alpha\mathbb{Z}) = \{0\}$ and j is not essential.

Let $\mathcal{A}$ be the collection of submodules T of N such that $M \cap T = \{0\}$. Then $\mathcal{A} \neq \emptyset$ since clearly $\{0\} \in \mathcal{A}$. Now $\mathcal{A}$ is inductively ordered; for, let $N_1 \subseteq N_2 \subseteq N_3 \subseteq \ldots$ be a chain of submodules in $\mathcal{A}$. Then $\bigcup_{i \geq 1} N_i = \sum_{i \geq 1} N_i$ is a submodule of N, and

$$M \cap \bigcup_{i \geq 1} N_i = \bigcup_{i \geq 1} (M \cap N_i) = \bigcup_{i \geq 1} \{0\} = \{0\}$$

so $\bigcup_{i \geq 1} N_i \in \mathcal{A}$. Applying Zorn's axiom we can therefore assert the existence of a submodule A of N that is maximal in $\mathcal{A}$.

Consider now the composite morphism

$$M \xrightarrow{\ i\ } N \xrightarrow{\ \natural\ } N/A.$$

That this is a monomorphism follows from the fact that if $x \in \operatorname{Ker} \natural \circ i$ then $\natural[i(x)] = 0$ and so $x = i(x) \in \operatorname{Ker} \natural = A$. Thus $x \in M \cap A = \{0\}$ and so $x = 0$.

To see that $\natural \circ i$ is essential, let G be a non-zero submodule of N/A. Then by the correspondence theorem we have $G = X/A$ where X is a non-zero submodule of N with $A \subseteq X$; and, indeed, $A \subset X$ since G is non-zero. Now if we had $(\natural \circ i)^{\leftarrow}(G) = \{0\}$ then we would have $\{0\} = i^{\leftarrow}[\natural^{\leftarrow}(X/A)] = X \cap M$ whence $X \in \mathcal{A}$ in contradiction to the maximality of A. Thus $(\natural \circ i)^{\leftarrow}(G) \neq \{0\}$ and so $\natural \circ i$ is essential.

4.9 (a) We have that

$$\begin{aligned} x \in f^{\rightarrow}(\operatorname{Ker} g \circ f) &\iff x = f(y) \text{ where } g[f(y)] = 0 \\ &\iff x = f(y) \text{ and } g(x) = 0 \\ &\iff x \in \operatorname{Im} f \cap \operatorname{Ker} g. \end{aligned}$$

(b) Likewise,

$$\begin{aligned} x \in g^{\leftarrow}(\operatorname{Im} g \circ f) &\iff (\exists y) \quad g(x) = g[f(y)] \\ &\iff (\exists y) \quad x - f(y) \in \operatorname{Ker} g \\ &\iff x \in \operatorname{Im} f + \operatorname{Ker} g. \end{aligned}$$

4.10 (a)$\Rightarrow$(b) : If $h \circ f = g$ then

$$x \in \operatorname{Ker} f \Rightarrow g(x) = h[f(x)] = h(0) = 0 \Rightarrow x \in \operatorname{Ker} g.$$

(b)$\Rightarrow$(a) : If $\operatorname{Ker} f \subseteq \operatorname{Ker} g$ then for $x, y \in A$ we have

$$f(x) = f(y) \Rightarrow x - y \in \operatorname{Ker} f \subseteq \operatorname{Ker} g \Rightarrow g(x) = g(y).$$

Since f is surjective, we can therefore define a mapping $h : B \to C$ by the prescription $h[f(x)] = g(x)$. Clearly then $h \circ f = g$, and h is an R–morphism since

$$\begin{aligned} h[f(x) + f(y)] &= h[f(x+y)] = g(x+y) \\ &= g(x) + g(y) = h[f(x)] + h[f(y)], \end{aligned}$$

$$h[\lambda f(x)] = h[f(\lambda x)] = g(\lambda x) = \lambda g(x) = \lambda h[f(x)].$$

If such an R–morphism h exists and if $k : B \to C$ is also an R–morphism with $k \circ f = g$ then $k \circ f = h \circ f$ whence $k = h$ (since f, being surjective, is right cancellable). Thus h is unique. Also,

$$\begin{aligned} \operatorname{Ker} h = \{0\} &\iff (g(x) = 0 \Rightarrow f(x) = 0) \\ &\iff \operatorname{Ker} g \subseteq \operatorname{Ker} f \\ &\iff \operatorname{Ker} g = \operatorname{Ker} f. \end{aligned}$$

Consider now the diagram

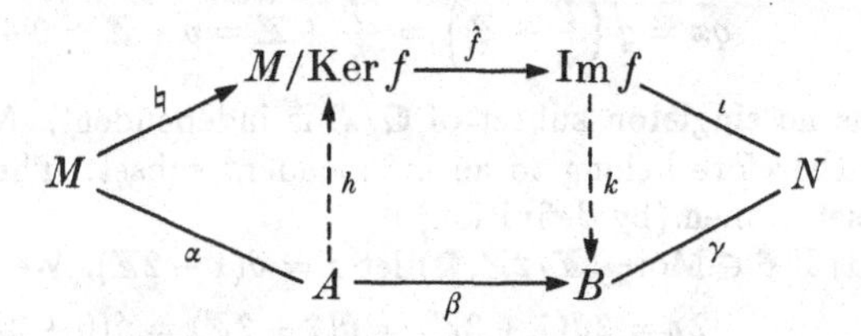

in which $\iota \circ \hat{f} \circ \natural$ is the canonical decomposition of f and α, β, γ are respectively an epimorphism, an isomorphism, and a monomorphism.

Since $f = \gamma \circ \beta \circ \alpha$ and since γ, β are each monomorphisms we have $f(x) = 0$ if and only if $\alpha(x) = 0$ so $\operatorname{Ker} \alpha = \operatorname{Ker} f = \operatorname{Ker} \natural$. By the previous part of the question it follows that there is a unique monomorphism $h : A \to M/\operatorname{Ker} f$ such that $h \circ \alpha = \natural$. Since $\natural$ is surjective it follows that so also is h. Thus h is an isomorphism.

Arguing dually, we see that $\operatorname{Im} \gamma = \operatorname{Im} f$ and there is a unique epimorphism $k : \operatorname{Im} f \to B$ such that $\gamma \circ k = \iota$. Since ι is injective so is k. Thus k is an isomorphism.

4.11 (a) Let $\{\frac{a_1}{a_1'}, \ldots, \frac{a_n}{a_n'}\}$ be a finite subset of $\mathbb{Q}$ and let $b = \prod_{i=1}^n a_i'$. Let p be a prime such that p does not divide b. Then $\frac{1}{p}$ is not in the submodule generated by $\{\frac{a_1}{a_1'}, \ldots, \frac{a_n}{a_n'}\}$. In fact, suppose that it were : then we would have, for some $x_1, \ldots, x_n \in \mathbb{Z}$,

$$\frac{1}{p} = \sum_{i=1}^{n} x_i \frac{a_i}{a_i'} = \frac{z}{b}$$

for some $z \in \mathbb{Z}$, so that $b = pz$ and we have the contradiction $p|b$. Thus $\mathbb{Q}$ is not finitely generated.

(b) If p, q are distinct primes then $\frac{1}{p} + \mathbb{Z} \neq \frac{1}{q} + \mathbb{Z}$, since otherwise $\frac{1}{p} - \frac{1}{q} \in \mathbb{Z}$ and so $q - p = pqn$ for some $n \in \mathbb{Z}$ whence, for example, $q = p(1 + qn)$ and the contradiction $p|q$. Thus we have that if $p_1, p_2, p_3, \ldots$ is the sequence of distinct primes then the elements

$$\frac{1}{p_1} + \mathbb{Z},\ \frac{1}{p_2} + \mathbb{Z},\ \frac{1}{p_3} + \mathbb{Z},\ \ldots$$

are distinct elements of $\mathbb{Q}/\mathbb{Z}$, so $\mathbb{Q}/\mathbb{Z}$ is infinite.

(c) For every $x \in \mathbb{Q}/\mathbb{Z}$ there exists $n \in \mathbb{Z}$ such that $n \neq 0$ and $nx = 0$. In fact, let $x = \frac{p}{q} + \mathbb{Z}$; then $q \neq 0$ and

$$qx = q\left(\frac{p}{q} + \mathbb{Z}\right) = \frac{qp}{q} + \mathbb{Z} = p + \mathbb{Z} = 0 + \mathbb{Z} = 0.$$

Thus no singleton subset of $\mathbb{Q}/\mathbb{Z}$ is independent. No element of $\mathbb{Q}/\mathbb{Z}$ can therefore belong to an independent subset. The only independent subset is then (by definition) $\emptyset$.

(d) If $\vartheta \in \mathrm{Mor}_{\mathbb{Z}}(\mathbb{Z}/2\mathbb{Z}, \mathbb{Q})$ let $x = \vartheta(1 + 2\mathbb{Z})$. We have

$$2x = 2\vartheta(1 + 2\mathbb{Z}) = \vartheta(2 + 2\mathbb{Z}) = \vartheta(0 + 2\mathbb{Z}) = 0$$

whence $x = 0$ and consequently $\vartheta = 0$.

(e) Let $\vartheta \in \mathrm{Mor}_{\mathbb{Z}}(\mathbb{Q}, \mathbb{Z})$ and suppose that $\vartheta(1) \neq 0$. Then for every non-zero $r \in \mathbb{Z}$ we have $\vartheta(1) = r\vartheta(\frac{1}{r})$ and so $r|\vartheta(1)$. Since $\vartheta(1)$ has only finitely many divisors, we deduce that we must have $\vartheta(1) = 0$. Consequently, for all $p, q \in \mathbb{Z}$ with $p, q \neq 0$ we have

$$0 = p\vartheta(1) = p\vartheta\left(\frac{q}{q}\right) = pq\vartheta\left(\frac{1}{q}\right) = q\vartheta\left(\frac{p}{q}\right)$$

whence $\vartheta(\frac{p}{q}) = 0$ and so ϑ is the zero map.

4.12 Consider $f : R \to M = Rx$ given by $f(\lambda) = \lambda x$. Clearly, f is an R-morphism with $\operatorname{Ker} f = \operatorname{Ann}_R(x)$. By the first isomorphism theorem,

$$M = \operatorname{Im} f \simeq R/\operatorname{Ker} f = R/\operatorname{Ann}_R(x).$$

That ϑ is well defined follows from the fact that

$$\begin{aligned} x + m\mathbb{Z} = y + m\mathbb{Z} &\Rightarrow x - y \in m\mathbb{Z} \\ &\Rightarrow nx - ny \in nm\mathbb{Z} \\ &\Rightarrow \vartheta(x + m\mathbb{Z}) = \vartheta(y + m\mathbb{Z}). \end{aligned}$$

Clearly, ϑ is a $\mathbb{Z}$-morphism.

Let $g \in \operatorname{Mor}_{\mathbb{Z}}(\mathbb{Z}/m\mathbb{Z}, \mathbb{Z}/nm\mathbb{Z})$ and let $g(1 + m\mathbb{Z}) = t + nm\mathbb{Z}$. Then

$$\begin{aligned} 0 + nm\mathbb{Z} = g(0 + m\mathbb{Z}) &= g(m + m\mathbb{Z}) \\ &= mg(1 + m\mathbb{Z}) \\ &= m(t + nm\mathbb{Z}) \\ &= mt + nm\mathbb{Z} \end{aligned}$$

so $mt \in nm\mathbb{Z}$ and consequently $t \in n\mathbb{Z}$, say $t = nr$. Thus $g(1 + n\mathbb{Z}) = nr + nm\mathbb{Z}$ and so, for every $x \in \mathbb{Z}$,

$$\begin{aligned} g(x + m\mathbb{Z}) = xg(1 + m\mathbb{Z}) &= xnr + nm\mathbb{Z} \\ &= r(nx + nm\mathbb{Z}) \\ &= r\vartheta(x + m\mathbb{Z}) \end{aligned}$$

whence we have $g = r\vartheta$. Thus $\operatorname{Mor}_{\mathbb{Z}}(\mathbb{Z}/m\mathbb{Z}, \mathbb{Z}/nm\mathbb{Z})$ is generated by $\{\vartheta\}$. Now

$$\begin{aligned} \lambda \in \operatorname{Ann}_{\mathbb{Z}}(\vartheta) \iff \lambda\vartheta = 0 &\iff (\lambda\vartheta)(1 + m\mathbb{Z}) = 0 + nm\mathbb{Z} \\ &\iff \lambda n + nm\mathbb{Z} = 0 + nm\mathbb{Z} \\ &\iff \lambda n \in nm\mathbb{Z} \\ &\iff \lambda \in m\mathbb{Z}, \end{aligned}$$

so $\operatorname{Ann}_{\mathbb{Z}}(\vartheta) = m\mathbb{Z}$. By the first part of the question we deduce that

$$\operatorname{Mor}_{\mathbb{Z}}(\mathbb{Z}/m\mathbb{Z}, \mathbb{Z}/nm\mathbb{Z}) \simeq \mathbb{Z}/\operatorname{Ann}_{\mathbb{Z}}(\vartheta) = \mathbb{Z}/m\mathbb{Z}.$$

4.13 Define $\vartheta : A \cap B \to A \times B$ by $\vartheta(x) = (x, x)$, and $\pi : A \times B \to A + B$ by $\pi(x, y) = x - y$. Clearly, ϑ is a monomorphism, and π is an epimorphism since $\pi(a, -b) = a + b$. Now $\operatorname{Im} \vartheta = \{(x, x) \mid x \in A \cap B\}$ and

$$\operatorname{Ker} \pi = \{(x, y) \mid x \in A, y \in B, x - y = 0\} = \{(x, x) \mid x \in A \cap B\},$$

so the sequence

$$0 \longrightarrow A \cap B \xrightarrow{\ \vartheta\ } A \times B \xrightarrow{\ \pi\ } A + B \longrightarrow 0$$

is exact.

4.14 (a) By commutativity we have

$$\alpha = \pi_1 \circ \alpha_2 = g_2 \circ \pi_2 \circ \iota_2 = g \circ 0 = 0.$$

Similarly, $\beta = 0$.

(b) Since $\pi_1 \circ \iota_2 = \alpha = 0$ we have $\operatorname{Im} \iota_2 \subseteq \operatorname{Ker} \pi_1 = \operatorname{Im} \iota_1$ so there is a unique R-morphism $k : A_2' \to A_1'$ such that $\iota_1 \circ k = \iota_2$. It follows that $\iota_1 \circ k \circ f = \iota_2 \circ f = \iota_1$ whence, ι_1 being injective (and hence left cancellable), $k \circ f = \mathrm{id}_{A_1'}$.

Since $\pi_2 \circ \iota_1 = \beta = 0$ we have similarly a unique R-morphism $k' : A_1' \to A_2'$ such that $\iota_2 \circ k' = \iota_1$. Then $k \circ k' = \mathrm{id}_{A_1'}$.

Now $\iota_2 \circ k' \circ k = \iota_1 \circ k = \iota_2$ so, ι_2 being left cancellable, $k' \circ k = \mathrm{id}_{A_2'}$. Thus we see that k is an isomorphism with $k^{-1} = k'$. Since $k \circ f = \mathrm{id}_{A_1'}$ we deduce that $f = k^{-1}$, whence f is an isomorphism. Similarly, so is g.

4.15 The first part is a special case of question 4.10, but we include the details.

(a)⇒(b) : If $f_\star \circ \natural_A = f$ then since $\operatorname{Ker} \natural_A = A$ we have

$$(\forall a \in A) \qquad f(a) = f_\star[\natural_A(a)] = f_\star(0) = 0$$

so $A \subseteq \operatorname{Ker} f$.

(b)⇒(a) : If $A \subseteq \operatorname{Ker} f$ then we have

$$x + A = y + A \Rightarrow x - y \in A \subseteq \operatorname{Ker} f \Rightarrow f(x) = f(y)$$

so we can define a mapping $f_\star : M/A \to N$ by setting $f_\star(x + A) = f(x)$. Clearly, $f_\star$ is an R-morphism, and $f_\star \circ \natural_A = f$.

That $f_\star$ is unique follows from the fact that $\natural_A$ is surjective and hence right cancellable.

Now $f_\star$ is a monomorphism if and only if

$$f(x) = 0 \Longrightarrow x + A = 0 + A = A$$

i.e. if and only if $\operatorname{Ker} f \subseteq A$, whence the result follows since the reverse inclusion holds by hypothesis.

Consider the sequence

$$0 \longrightarrow M/(A \cap B) \xrightarrow{\ \alpha\ } M/A \times M/B \xrightarrow{\ \beta\ } M/(A + B) \longrightarrow 0.$$

Let β be given by $\beta(x + A, y + B) = x - y + A + B$. Then β is an R-epimorphism; for example, $\beta(z + A, 0 + B) = z + A + B$.

As for α, consider first the mapping $f : M \to M/A \times M/B$ given by $f(x) = (x + A, x + B)$. This is an R-morphism with $\operatorname{Ker} f = A \cap B$ so, by the first part of the question, there is a (unique) monomorphism $\alpha : M/(A \cap B) \to M/A \times M/B$ such that $\alpha \circ \natural_{A\cap B} = f$.

To establish exactness, we must show that $\operatorname{Im} \alpha = \operatorname{Ker} \beta$. Now

$$\operatorname{Im} \alpha = \{(x + A, x + B) \mid x \in M\};$$

and

$$\operatorname{Ker} \beta = \{(x + A, y + B) \mid x - y \in A + B\}.$$

But we have that

$$\begin{aligned} x - y \in A + B &\iff (\exists a \in A)(\exists b \in B)\ x - y = a + b \\ &\iff (\exists a \in A)(\exists b \in B)\ x - a = y + b = z \text{ say} \\ &\iff (\exists z \in M)\ x + A = z + A,\ y + B = z + B. \end{aligned}$$

Hence we see that $\operatorname{Im} \alpha = \operatorname{Ker} \beta$.

For the last part, take $M = A + B$ to get the sequence

$$0 \longrightarrow (A+B)/(A \cap B) \xrightarrow{\ \alpha\ } (A+B)/A \times (A+B)/B \longrightarrow 0 \longrightarrow 0$$

the exactness of which shows that α is an isomorphism.

4.16 Suppose that the top row is exact. Then

$$g' \circ f' \circ \alpha = \gamma \circ g \circ f = \gamma \circ 0 = 0$$

so, α being right cancellable, $g' \circ f' = 0$. It now follows that $\operatorname{Im} f' \subseteq \operatorname{Ker} g'$.

To obtain the reverse inclusion, let $b' \in \operatorname{Ker} g'$. Since β is an isomorphism, there is a unique $b \in B$ such that $b' = \beta(b)$. Then

$$0 = g'(b') = g'[\beta(b)] = \gamma[g(b)]$$

so, γ being an isomorphism, $g(b) = 0$ whence $b \in \operatorname{Ker} g = \operatorname{Im} f$ so $b = f(a)$ for some $a \in A$. Then

$$b' = \beta(b) = \beta[f(a)] = f'[\alpha(a)] \in \operatorname{Im} f'.$$

Thus the bottom row is exact.

4.17 $g \circ \alpha \circ f' = g \circ f \circ \alpha' = 0 \circ \alpha' = 0$ so $\operatorname{Ker} g' = \operatorname{Im} f' \subseteq \operatorname{Ker} g \circ \alpha$. Since g' is surjective it follows that there is a unique R-morphism $\alpha'' : C' \to C$ such that $\alpha'' \circ g' = g \circ \alpha$. Similarly, $g'' \circ \beta \circ f = 0$ so $\operatorname{Ker} g = \operatorname{Im} f \subseteq \operatorname{Ker} g'' \circ \beta$ and there is a unique R-morphism $\beta'' : C \to C''$ such that $\beta'' \circ g = g'' \circ \beta$.

To show that the thus completed bottom row is exact, we have to show that

(1) α'' is injective;
(2) β'' is surjective;
(3) $\operatorname{Im} \alpha'' = \operatorname{Ker} \beta''$.

This we do by the usual process of 'diagram chasing'.

(1) Let $c' \in \operatorname{Ker} \alpha''$. Since g' is surjective there exists $b' \in B'$ with $g'(b') = c'$. Then

$$0 = \alpha''(c') = \alpha''[g'(b')] = g[\alpha(b')]$$

so $\alpha(b') \in \operatorname{Ker} g = \operatorname{Im} f$ so $\alpha(b') = f(a)$ for some $a \in A$. Since then

$$0 = \beta[\alpha(b')] = \beta[f(a)] = f''[\beta'(a)]$$

we have $\beta'(a) \in \operatorname{Ker} f'' = \{0\}$ so $a \in \operatorname{Ker} \beta' = \operatorname{Im} \alpha'$, say $a = \alpha'(a')$. Consequently

$$\alpha(b') = f(a) = f[\alpha'(a')] = \alpha[f'(a')],$$

giving $b' = f'(a')$ since α is monic. It now follows that

$$c' = g'(b') = g'[f'(a')] = 0$$

and so $\operatorname{Ker} \alpha'' = \{0\}$ as required.

(2) Since $\beta'' \circ g = g'' \circ \beta$ and since g'', β are each surjective we see that $\beta'' \circ g$ is surjective, whence so also is β''.

(3) $\beta'' \circ \alpha'' \circ g' = g'' \circ \beta \circ \alpha = g'' \circ 0 = 0$ so, g' being surjective (and hence right cancellable), $\beta'' \circ \alpha'' = 0$ whence $\operatorname{Im} \alpha'' \subseteq \operatorname{Ker} \beta''$.

To obtain the reverse inclusion, let $c \in \operatorname{Ker} \beta''$. Then, since g is surjective, $c = g(b)$ for some $b \in B$ so

$$0 = \beta''(c) = \beta''[g(b)] = g''[\beta(b)]$$

whence $\beta(b) \in \operatorname{Ker} g'' = \operatorname{Im} f''$. Thus $\beta(b) = f''(a'')$ for some $a'' \in A''$ and, since β' is surjective, $a'' = \beta'(a)$ for some $a \in A$. Thus

$$\beta(b) = f''[\beta'(a)] = \beta[f(a)]$$

giving $b - f(a) \in \operatorname{Ker}\beta = \operatorname{Im}\alpha$. Thus, for some $b' \in B'$, we have $b - f(a) = \alpha(b')$ and it follows that

$$\begin{aligned} c = g(b) = g[f(a) + \alpha(b')] &= g[f(a)] + g[\alpha(b')] \\ &= g[\alpha(b')] \qquad \text{since } g \circ f = 0 \\ &= \alpha''[g'(b')] \\ &\in \operatorname{Im}\alpha'' \end{aligned}$$

so $\operatorname{Ker}\beta'' \subseteq \operatorname{Im}\alpha''$.

4.18 Clearly, $Rx^{n+1} = Rx.x^n \subseteq Rx^n$ so we have the descending chain

$$R \supseteq Rx \supseteq Rx^2 \supseteq \cdots \supseteq Rx^n \supseteq Rx^{n+1} \supseteq \cdots$$

of submodules of R.

Consider the mapping $f : R \to Rx^n/Rx^{n+1}$ given by the prescription $f(r) = rx^n + Rx^{n+1}$. It is clear that f is an R-epimorphism. Now

$$\begin{aligned} \operatorname{Ker} f &= \{r \in R \mid rx^n \in Rx^{n+1}\} \\ &= \{r \in R \mid (\exists t \in R)\ rx^n = tx^{n+1}\} \\ &= \{r \in R \mid (\exists t \in R)\ (r - tx)x^n = 0\} \\ &= \{r \in R \mid (\exists t \in R)\ r - tx = 0\} \\ &= Rx, \end{aligned}$$

the penultimate equality resulting from the fact that $x \neq 0$ and R has no zero divisors. It now follows by the first isomorphism theorem that

$$Rx^n/Rx^{n+1} = \operatorname{Im} f \simeq R/\operatorname{Ker} f = R/Rx.$$

4.19 (a) The $\mathbb{Z}$-module $\mathbb{Z}$ satisfies the ascending chain condition but not the descending chain condition. For, every submodule of $\mathbb{Z}$ is finitely generated (in fact by a singleton); and an infinite descending chain of submodules is

$$2\mathbb{Z} \supset 4\mathbb{Z} \supset 8\mathbb{Z} \supset 16\mathbb{Z} \supset \ldots.$$

(b) The $\mathbb{Z}$-module $\mathbb{Z}_m$ satisfies both chain conditions since it is finite.

(c) This is similar to (b).

(d) The $\mathbb{Q}$-module $\mathbb{Q}$ is a vector space of dimension 1 over $\mathbb{Q}$. Any submodule of this is therefore a subspace and so is either $\{0\}$ or $\mathbb{Q}$. Thus both chain conditions are satisfied.

(e) As a $\mathbb{Z}$-module, $\mathbb{Q}$ satisfies neither chain condition. To see this, let the submodule generated by $t \in \mathbb{Q}$ be given by $(t) = \{nt \mid n \in \mathbb{Z}\}$. Then

$$(2) \supset (4) \supset (8) \supset (16) \supset \cdots$$

is an infinite descending chain of submodules, and

$$(\tfrac{1}{2}) \subset (\tfrac{1}{4}) \subset (\tfrac{1}{8}) \subset (\tfrac{1}{16}) \subset \cdots$$

is an infinite ascending chain of submodules.

(f) As a $\mathbb{Q}$-module, $\mathbb{Q}[X]$ satisfies neither chain condition. To see this, let $(t) = t\mathbb{Q}$ be the submodule generated by $t \in \mathbb{Q}[X]$. Then

$$(X, X^2, X^3, X^4, \ldots) \supset (X^2, X^3, X^4, \ldots) \supset (X^3, X^4, \ldots) \supset \cdots$$

is an infinite descending chain of submodules, and

$$(X) \subset (X, X^2) \subset (X, X^2, X^3) \subset (X, X^2, X^3, X^4) \subset \cdots$$

is an infinite ascending chain of submodules.

(g) As a $\mathbb{Q}[X]$-module, $\mathbb{Q}[X]$ satisfies the ascending chain condition but not the descending chain condition. In fact

$$(X) \supset (X^2) \supset (X^3) \supset (X^4) \supset \cdots$$

is an infinite descending chain of submodules, and every submodule is finitely generated (since a submodule of $\mathbb{Q}[X]$ as a $\mathbb{Q}[X]$-module is an ideal which, by the division algorithm, is generated by the monic polynomial of least degree that it contains).

(h) By the correspondence theorem the submodules of $\mathbb{Q}[X]/M$ are of the form T/M where T is a submodule of $\mathbb{Q}[X]$ that contains M. But the only such submodules T are of the form $\big(f(X)\big)$ where $f(X)$ divides X^5, and there are finitely many of these submodules. Hence $\mathbb{Q}[X]/M$ as a $\mathbb{Q}[X]$-module satisfies both chain conditions.

(i) As in (h).

4.20 Every tower of submodules can be refined to a Jordan–Hölder tower. In particular, the tower $\{0\} \subseteq N \subseteq M$ can be so refined.

4.21 By the previous question there is a Jordan–Hölder tower that 'passes through' N, say

$$M = M_0 \supset M_1 \supset \cdots \supset M_n = N \supset M_{n+1} \supset \cdots \supset M_{n+t} = 0$$

which is of height $n+t$. Since

$$M/N = M_0/N \supset M_1/N \supset \cdots \supset M_n/N = N/N = 0$$

is then a Jordan–Hölder tower (of height n) for M/N and

$$N = M_n \supset M_{n+1} \supset \cdots \supset M_{n+t} = 0$$

is a Jordan–Hölder tower (of height t) for N, we deduce that

$$h(M) = n + t = h(M/N) + h(N).$$

For the last part observe that

$$N = M \iff M/N = 0 \iff h(M/N) = 0 \iff h(M) = h(N).$$

4.22 Since M is of finite height and $\operatorname{Ker} f$ is a submodule of M it follows that $\operatorname{Ker} f$ is of finite height. Since $\operatorname{Im} f \simeq M/\operatorname{Ker} f$ and since $M/\operatorname{Ker} f$ is also of finite height (for quotient modules inherit chain conditions), it follows that $\operatorname{Im} f$ is of finite height. It now follows by the previous question that

$$h(\operatorname{Im} f) + h(\operatorname{Ker} f) = h(M/\operatorname{Ker} f) + h(\operatorname{Ker} f) = h(M).$$

4.23 The proof is by induction. If $n = 1$ then the sequence

$$0 \longrightarrow M_1 \longrightarrow 0$$

can only be exact if $M_1 = 0$, in which case the result is trivial. When $n = 2$ the exactness of

$$0 \longrightarrow M_1 \xrightarrow{f} M_2 \longrightarrow 0$$

requires f to be an isomorphism, in which case $h(M_1) = h(M_2)$. Suppose then that $n > 2$ and that the result holds for all exact sequences of the given type of length less than or equal to $n-1$. Consider the diagram

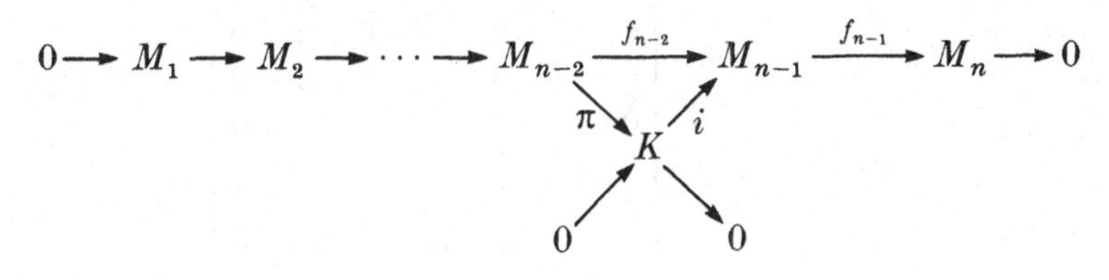

in which the row is exact, $K = \text{Ker}\, f_{n-1} = \text{Im}\, f_{n-2}$, π is the epimorphism induced by f_{n-2}, and i is the canonical inclusion.

The sequence

$$0 \longrightarrow M_1 \longrightarrow M_2 \longrightarrow \cdots \longrightarrow M_{n-2} \xrightarrow{\pi} K \longrightarrow 0$$

is exact and so, by the inductive hypothesis,

$$\sum_{k=1}^{n-2}(-1)^k h(M_k) + (-1)^{n-1}h(K) = 0.$$

Also, the sequence

$$0 \longrightarrow K \longrightarrow M_{n-1} \xrightarrow{f_{n-1}} M_n \longrightarrow 0$$

is exact and so, again by the inductive hypothesis,

$$-h(K) + h(M_{n-1}) - h(M_n) = 0.$$

It follows from this that

$$(-1)^{n-1}h(K) = (-1)^{n-1}[h(M_{n-1}) - h(M_n)] = 0.$$

Substituting for $(-1)^{n-1}h(K)$ in the previous equation now gives the result for n and completes the induction.

4.24 Consider $M = \mathbb{Z}_2 \oplus \mathbb{Z}_2$ and $N = \mathbb{Z}_3 \oplus \mathbb{Z}_3$. Clearly, $h(M) = h(N) = 2$. Also, every element of M is of order 2. If now $\vartheta \in \text{Mor}_{\mathbb{Z}}(M, N)$ then

$$(\forall x \in M) \qquad 0 = \vartheta(0) = \vartheta(x + x) = \vartheta(x) + \vartheta(x).$$

Since N has no elements of order 2 we deduce that $\vartheta(x) = 0$ for all $x \in M$ whence $\vartheta = 0$.

4.25 It is readily seen that $M_n E_i$ is the submodule of M_n consisting of those matrices of the form

$$\begin{bmatrix} 0 & \dots & 0 & a_{1i} & 0 & \dots & 0 \\ 0 & \dots & 0 & a_{2i} & 0 & \dots & 0 \\ 0 & \dots & 0 & a_{3i} & 0 & \dots & 0 \\ \vdots & & \vdots & \vdots & \vdots & & \vdots \\ 0 & \dots & 0 & a_{ni} & 0 & \dots & 0 \end{bmatrix}$$

and that B_i is the submodule of M_n consisting of those matrices of the form

$$\begin{bmatrix} a_{11} & a_{12} & \dots & a_{1i} & 0 & \dots & 0 \\ a_{21} & a_{22} & \dots & a_{2i} & 0 & \dots & 0 \\ a_{31} & a_{32} & \dots & a_{3i} & 0 & \dots & 0 \\ \vdots & \vdots & & \vdots & \vdots & & \vdots \\ a_{n1} & a_{n2} & \dots & a_{ni} & 0 & \dots & 0 \end{bmatrix}.$$

We thus have the tower

$$M_n = B_n \supset B_{n-1} \supset \cdots \supset B_1 \supset B_0 = \{0\}.$$

Now there are no submodules A such that $B_{i-1} \subset A \subset B_i$ and so, by the correspondence theorem, B_i/B_{i-1} is simple. This is therefore a Jordan–Hölder tower.

4.26 For the first part it suffices to prove that

$$N_i \cap \sum_{j \neq i} N_j = \{0\}$$

for $i = 1, \dots, n$. This follows from the fact that $N_i \subseteq M_i$ for every i and $M = \bigoplus_{i=1}^{n} M_i$, so that

$$N_i \cap \sum_{j \neq i} N_j \subseteq M_i \cap \sum_{j \neq i} M_j = \{0\}.$$

Consider now the mapping

$$\vartheta : M = \bigoplus_{i=1}^{n} M_i \longrightarrow \bigoplus_{i=1}^{n} M_i/N_i$$

given by the prescription

$$\vartheta(m_1, \dots, m_n) = (m_1 + N_1, \dots, m_n + N_n).$$

Clearly, ϑ is an R–morphism with

$$\operatorname{Ker} \vartheta = \{(m_1, \dots, m_n) \mid m_i \in N_i\} = \bigoplus_{i=1}^{n} N_i = N.$$

Then, by the first isomorphism theorem,

$$\bigoplus_{i=1}^{n} M_i/N_i = \operatorname{Im}\vartheta \simeq M/\operatorname{Ker}\vartheta = M/N.$$

4.27 Suppose that $M = \bigoplus_{i=1}^{n} A_i$ and let $\mathrm{pr}_i : M \to M_i$ and $\mathrm{in}_i : M_i \to M$ be the ith projection and injection respectively, i.e.

$$\mathrm{pr}_i(a_1,\ldots,a_n) = a_i,$$
$$\mathrm{in}_i(a_i) = (\underbrace{0,\ldots,0,a_i}_{i},0,\ldots,0).$$

Then it is clear that pr_i and in_i satisfy (1) and (2).

Conversely, suppose that $f_i : M \to A_i$ and $g_i : A_i \to M$ satisfy (1) and (2). Let X be any R-module and for each $i \in I$ let $h_i : A_i \to X$ be an R-morphism. Consider the diagram

$$\begin{array}{ccc} X & \xleftarrow{h_i} & A_i \\ & & f_i \downarrow\uparrow g_i \\ & & M \end{array}$$

Define $h : M \to X$ by

$$h = \sum_{j=1}^{n} h_j \circ g_j.$$

For every i we have

$$h \circ f_i = \sum_{j=1}^{n} h_j \circ g_j \circ f_i \overset{(1)}{=} h_i.$$

If, moreover, $t : M \to X$ is such that $t \circ f_i = h_i$ for every i then

$$t = t \circ \mathrm{id}_M \overset{(2)}{=} t \circ \sum_{j=1}^{n} f_j \circ g_j = \sum_{j=1}^{n} t \circ f_j \circ g_j = \sum_{j=1}^{n} h_j \circ g_j = h.$$

Thus M is the direct sum of $A_1,\ldots,A_n$.

We have that $M = M_1 \oplus M_2$ if and only if there exist R-morphisms

$$M_1 \underset{g_1}{\overset{f_1}{\rightleftarrows}} M \underset{f_2}{\overset{g_2}{\rightleftarrows}} M_2$$

such that

$$\text{(a) } g_1 \circ f_1 = \mathrm{id}_{M_1}; \qquad \text{(b) } g_1 \circ f_2 = 0;$$
$$\text{(c) } g_2 \circ f_2 = \mathrm{id}_{M_2}; \qquad \text{(d) } g_2 \circ f_1 = 0;$$
$$\text{(e) } f_1 \circ g_1 + f_2 \circ g_2 = \mathrm{id}_M .$$

Now if these conditions are satisfied then the sequence

$$0 \longrightarrow M_1 \xrightarrow{f_1} M \xrightarrow{g_2} M_2 \longrightarrow 0$$

is exact; for f_1 is injective by (a), g_2 is surjective by (c), $\mathrm{Im}\, f_1 \subseteq \mathrm{Ker}\, g_2$ by (d), and by (e) if $g_2(x) = 0$ then $x = (f_1 \circ g_1)(x)$ so that $\mathrm{Ker}\, g_2 \subseteq \mathrm{Im}\, f_1$. Moreover, the sequence splits by (a) and (c).

4.28 Note that R is free with basis $\{1\}$. If

$$0 \longrightarrow X \longrightarrow A \xrightarrow{f} R \longrightarrow 0$$

is exact then f is surjective so there exists $a \in A$ such that $f(a) = 1$. Choosing any such $a \in A$ define $g : R \to A$ by $g(1) = a$ and extend this to the whole of R by linearity. Then

$$(f \circ g)(1) = f(a) = 1$$

so $f \circ g$ and id_R coincide on the basis $\{1\}$. Hence $f \circ g = \mathrm{id}_R$ and g is a splitting morphism.

The sequence

$$0 \longrightarrow \mathbb{Z} \xrightarrow{\times 2} \mathbb{Z} \xrightarrow{\natural} \mathbb{Z}/2\mathbb{Z} \longrightarrow 0$$

is exact, but cannot be split since, as can be seen in a similar manner to part (d) of question 4.11, we have

$$\mathrm{Mor}_{\mathbb{Z}}(\mathbb{Z}/2\mathbb{Z}, \mathbb{Z}) = 0.$$

4.29 To say that $\mathrm{Ker}\, f$ is a direct summand of M and that $\mathrm{Im}\, f$ is a direct summand of N is equivalent to saying that in the following diagram (in which $i_2 \circ b \circ \natural_1$ is the canonical decomposition of f) both short exact sequences split :

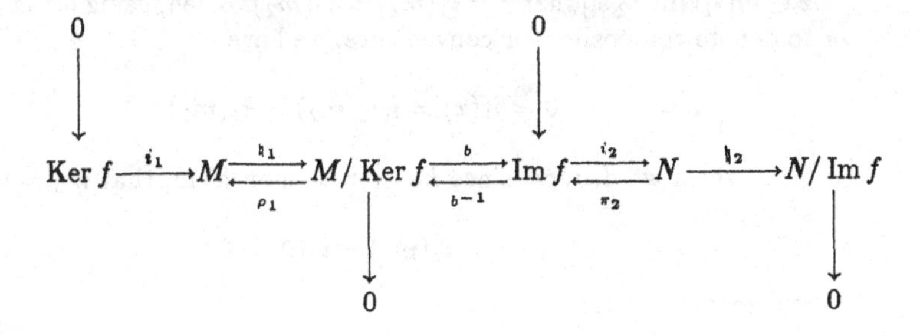

Let $g : N \to M$ be given by

$$g = \rho_1 \circ b^{-1} \circ \pi_2.$$

Then we have

$$\begin{aligned} f \circ g \circ f &= i_2 \circ b \circ \natural_1 \circ \underbrace{\rho_1 \circ b^{-1} \circ \pi_2 \circ i_2}_{} \circ b \circ \natural_1 \\ &= i_2 \circ b \circ b^{-1} \circ b \circ \natural_1 \\ &= i_2 \circ b \circ \natural_1 \\ &= f, \end{aligned}$$

where $\natural_1 \circ \rho_1 = \mathrm{id}$ and $\pi_2 \circ i_2 = \mathrm{id}$,

and so f is regular.

Conversely, suppose that f is regular. Then there exists $g : N \to M$ such that $f \circ g \circ f = f$. Using the canonical decomposition $f = i_2 \circ b \circ \natural_1$ this translates into

$$i_2 \circ b \circ \natural_1 \circ g \circ i_2 \circ b \circ \natural_1 = i_2 \circ b \circ \natural_1.$$

Using the fact that i_2 is left cancellable, that $\natural_1$ is right cancellable, and that b is cancellable on both sides, we deduce from this that, on the one hand,

$$b \circ \natural_1 \circ g \circ i_2 = \mathrm{id}_{\mathrm{Im}\, f}$$

and, on the other,

$$\natural_1 \circ g \circ i_2 \circ b = \mathrm{id}_{M/\mathrm{Ker}\, f}.$$

Thus $\pi_2 = b \circ \natural_1 \circ g$ and $\rho_1 = g \circ i_2 \circ b$ are splitting morphisms for the above short exact sequences and the result follows.

4.30 If $x \in \mathrm{Ker}\, j_1 \cap \mathrm{Ker}\, j_2$ then $j_1(x) = 0 = j_2(x)$ and, by the exactness, $x \in \mathrm{Im}\, i_1 \cap \mathrm{Im}\, i_2$ so that $x = i_1(m_1) = i_2(m_2)$. Then, using juxtaposition to denote composites for convenience, we have

$$0 = j_1(x) = j_1 i_2(m_2) = k_1(m_2)$$

from which we deduce, since k_1 is an isomorphism, that $m_2 = 0$. Consequently,

$$x = i_2(m_2) = i_2(0) = 0$$

and so $\mathrm{Ker}\, j_1 \cap \mathrm{Ker}\, j_2 = \{0\}$.

If we define $\bar{x} = i_1 k_2^{-1} j_2(x) + i_2 k_1^{-1} j_1(x)$ then

$$j_1(\bar{x}) = \underbrace{j_1 i_1}_{0} k_2^{-1} j_2(x) + \underbrace{j_1 i_2}_{k_1} k_1^{-1} j_1(x) = j_1(x)$$

$$j_2(\bar{x}) = \underbrace{j_2 i_1}_{k_2} k_2^{-1} j_2(x) + \underbrace{j_2 i_2}_{0} k_1^{-1} j_1(x) = j_2(x)$$

and hence $\bar{x} - x \in \operatorname{Ker} j_1 \cap \operatorname{Ker} j_2$. It now follows from the first part that $\bar{x} = x$.

The fact that $\bar{x} = x$ can be translated into

$$i_1 k_2^{-1} j_2 + i_2 k_1^{-1} j_1 = \mathrm{id}_M .$$

Thus we see that $M = \operatorname{Im} i_1 + \operatorname{Im} i_2$. Since also

$$\operatorname{Im} i_1 \cap \operatorname{Im} i_2 = \operatorname{Ker} j_1 \cap \operatorname{Ker} j_2 = \{0\},$$

it follows that $M = \operatorname{Im} i_1 \oplus \operatorname{Im} i_2$.

For the last part we observe that

$$\begin{aligned} h_1 k_1^{-1} \ell_1 + h_2 k_2^{-1} \ell_2 &= j_0 i_2 k_1^{-1} j_1 i_0 + j_0 i_1 k_2^{-1} j_2 i_0 \\ &= j_0 (i_2 k_1^{-1} j_1 + i_1 k_2^{-1} j_2) i_0 \\ &= j_0 \circ \mathrm{id}_M \circ i_0 \\ &= j_0 i_0 \\ &= 0. \end{aligned}$$

4.31 We have that

$$\begin{aligned} (m_\alpha)_{\alpha \in I} \in \operatorname{Im} f &\iff (\forall \alpha \in I)(\exists m_\alpha \in M_\alpha)\ f_\alpha(m_\alpha) = n_\alpha \\ &\iff (\forall \alpha \in I)\ n_\alpha \in \operatorname{Im} f_\alpha \\ &\iff (n_\alpha)_{i \in I} \in \bigoplus_{\alpha \in I} \operatorname{Im} f_\alpha; \\ (m_\alpha)_{\alpha \in I} \in \operatorname{Ker} f &\iff (\forall \alpha \in I)\ f_\alpha(m_\alpha) = 0 \\ &\iff (\forall \alpha \in I)\ m_\alpha \in \operatorname{Ker} f_\alpha \\ &\iff (m_\alpha)_{\alpha \in I} \in \bigoplus_{\alpha \in I} \operatorname{Ker} f_\alpha, \end{aligned}$$

so that $\operatorname{Im} f = \bigoplus_{\alpha \in I} \operatorname{Im} f_\alpha$ and $\operatorname{Ker} f = \bigoplus_{\alpha \in I} \operatorname{Ker} f_\alpha$.

If every sequence $L_\alpha \xrightarrow{g_\alpha} M_\alpha \xrightarrow{f_\alpha} N_\alpha$ is exact then for every $\alpha \in I$ we have $\operatorname{Im} g_\alpha = \operatorname{Ker} f_\alpha$ and hence

$$\operatorname{Im} g = \bigoplus_{\alpha \in I} \operatorname{Im} g_\alpha = \bigoplus_{\alpha \in I} \operatorname{Ker} f_\alpha = \operatorname{Ker} f.$$

Thus the sequence

$$\bigoplus_{\alpha \in I} L_\alpha \xrightarrow{g} \bigoplus_{\alpha \in I} M_\alpha \xrightarrow{f} \bigoplus_{\alpha \in I} N_\alpha$$

is also exact.

Conversely, suppose that the direct sum sequence is exact. Then we have the commutative diagrams (one for each α)

$$\begin{array}{ccccc} L_\alpha & \xrightarrow{f_\alpha} & M_\alpha & \xrightarrow{g_\alpha} & N_\alpha \\ \downarrow{\scriptstyle i_1} & & \downarrow{\scriptstyle i_2} & & \downarrow{\scriptstyle i_3} \\ \bigoplus L_\alpha & \xrightarrow[f]{} & \bigoplus M_\alpha & \xrightarrow[g]{} & \bigoplus N_\alpha \end{array}$$

in which the bottom row is exact and i_1, i_2, i_3 are the canonical injections. Now $g \circ f = 0$ gives $g \circ f \circ i_1 = 0$ and so, by commutativity, $i_3 \circ g_\alpha \circ f_\alpha = 0$. Since i_3 is monic, and so left cancellable, we deduce that $g_\alpha \circ f_\alpha = 0$ and hence that $\operatorname{Im} f_\alpha \subseteq \operatorname{Ker} g_\alpha$.

In order to establish the reverse inclusion, given $m_\alpha \in M_\alpha$ let $\overline{m}_\alpha$ denote the element of $\bigoplus_{\alpha \in I} M_\alpha$ whose α-component is m_α and all other components are 0. Then we have

$$\begin{aligned} m_\alpha \in \operatorname{Ker} g_\alpha &\Rightarrow \overline{m}_\alpha \in \operatorname{Ker} g \quad \text{(RH square commutative)} \\ &\Rightarrow \overline{m}_\alpha \in \operatorname{Im} f \quad \text{(bottom row exact)} \\ &\Rightarrow m_\alpha \in \operatorname{Im} f_\alpha \quad \text{(LH square commutative)} \end{aligned}$$

and so $\operatorname{Ker} g_\alpha \subseteq \operatorname{Im} f_\alpha$.

Thus we see that the top row is also exact.

4.32 Note first that, since R is commutative, $\operatorname{Mor}_R(M, N)$ is an R-module. Also, $L_j \neq \emptyset$ since it clearly contains the zero morphism from M to N. To show that L_j is an R-module, it therefore suffices to show that it is a submodule of $\operatorname{Mor}_R(M, N)$; and for this it suffices to prove that if

$f, g \in L_j$ then $f + g \in L_j$, and if $f \in L_j, \lambda \in R$ then $\lambda f \in L_j$. For the first of these, observe that

$$\operatorname{Ker} f \cap \operatorname{Ker} g \subseteq \operatorname{Ker}(f+g),$$

since $f(x) = 0 = g(x)$ implies $(f+g)(x) = f(x) + g(x) = 0$; and for the second, observe that

$$\operatorname{Ker} f \subseteq \operatorname{Ker} \lambda f,$$

since $f(x) = 0$ implies $(\lambda f)(x) = \lambda f(x) = \lambda 0 = 0$.

As for the stated isomorphism, note that by applying the first isomorphism theorem to $\mathrm{pr}_j : M = \bigoplus_{i=1}^{n} A_i \to A_j$ we obtain

$$A_j \simeq M / \bigoplus_{i \neq j} A_i.$$

Let $\vartheta_j : A_j \to M / \bigoplus_{i \neq j} A_i$ be the R-isomorphism given by

$$\vartheta_j(a_j) = a_j + \bigoplus_{i \neq j} A_i.$$

If $f \in L_j$ then $\bigoplus_{i \neq j} A_i \subseteq \operatorname{Ker} f$ so there is a unique R-morphism $f_\star : M / \bigoplus_{i \neq j} A_i \to N$ such that $f_\star \circ \natural = f$ in the diagram

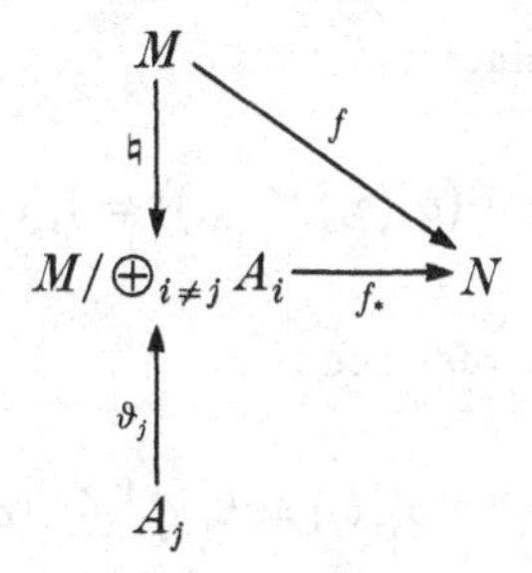

Then clearly $f_\star \circ \vartheta_j \in \operatorname{Mor}_R(A_j, N)$. Consider the mapping described by $f \mapsto f_\star \circ \vartheta_j$. This is injective, since ϑ_j is an isomorphism and $f_\star$ is unique; and it is an R-morphism, since the uniqueness of $f_\star$ gives $(f+g)_\star = f_\star + g_\star$ and $(\lambda f)_\star = \lambda f_\star$. It remains, therefore, to show that this map is surjective. Now given any $h \in \operatorname{Mor}_R(A_j, N)$ we have $h \circ \mathrm{pr}_j \in \operatorname{Mor}_R(M, N)$. Since clearly

$$(h \circ \mathrm{pr}_j)^{\rightarrow}\big(\bigoplus_{i \neq j} A_i\big) = 0,$$

we have $h \circ \mathrm{pr}_j \in L_j$. Moreover, for all $a_j \in A_j$ we have

$$\begin{aligned}[(h \circ \mathrm{pr}_j)_\star \circ \vartheta_j](a_j) &= (h \circ \mathrm{pr}_j)_\star\big(a_j + \bigoplus_{i \neq j} A_i\big) \\ &= (h \circ \mathrm{pr}_j)(a_j) \\ &= h(a_j)\end{aligned}$$

whence we see that $h = (h \circ \mathrm{pr}_j)_\star \circ \vartheta_j$.

4.33 First we show that the sequence is *semi-exact* in the sense that the composite of successive morphisms is zero; this will show that the image of the 'input' morphism at each stage is contained in the kernel of the 'output' morphism, for

$$f \circ g = 0 \iff \operatorname{Im} g \subseteq \operatorname{Ker} f.$$

We shall find it convenient to mix notation and use juxtaposition to denote composition.

(1) $\varphi_i \circ h_{i-1}\gamma_{i-1}^{-1}g'_{i-1} = 0$ follows from the given exactness, commutativity, and

$$\alpha_i h_{i-1}\gamma_{i-1}^{-1}g'_{i-1} = h'_{i-1}\gamma_{i-1}\gamma_{i-1}^{-1}g'_{i-1} = h'_{i-1}g'_{i-1} = 0;$$
$$f_i h_{i-1}\gamma_{i-1}^{-1}g'_{i-1} = 0.$$

(2) $\vartheta_i \circ \varphi_i = 0$ since

$$\vartheta_i[\varphi_i(a_i)] = \vartheta_i\big(\alpha_i(a_i), f_i(a_i)\big) = f'_i[\alpha_i(a_i)] - \beta_i[f_i(a_i)] = 0.$$

(3) $h_i\gamma_i^{-1}g'_i \circ \vartheta_i = 0$ since

$$\begin{aligned}(h_i\gamma_i^{-1}g'_i\vartheta_i)(a'_i, b_i) &= h_i\gamma_i^{-1}g'_if'_i(a'_i) - h_i\gamma_i^{-1}g'_i\beta_i(b_i)\\ &= 0 - h_i\gamma_i^{-1}\gamma_i g_i(b_i)\\ &= 0.\end{aligned}$$

To show that the sequence is exact at B'_i, it now suffices to observe that

$$\begin{aligned}b'_i \in \operatorname{Ker} h_i\gamma_i^{-1}g'_i &\Rightarrow h_i\gamma_i^{-1}g'_i(b'_i) = 0\\ &\Rightarrow \gamma_i^{-1}g'_i(b'_i) \in \operatorname{Ker} h_i = \operatorname{Im} g_i\\ &\Rightarrow \gamma_i^{-1}g'_i(b'_i) = g_i(b_i)\\ &\Rightarrow g'_i(b'_i) = \gamma_i g_i(b_i) = g'_i\beta_i(b_i)\\ &\Rightarrow b'_i - \beta_i(b_i) \in \operatorname{Ker} g'_i = \operatorname{Im} f'_i\\ &\Rightarrow b'_i = \beta_i(b_i) + f'_i(a'_i) = f'_i(a'_i) - \beta_i(-b_i) \in \operatorname{Im} \vartheta_i.\end{aligned}$$

As for exactness at $A'_i \oplus B_i$, first observe that

$$
\begin{aligned}
(a'_i, b_i) \in \operatorname{Ker} \vartheta_i & \\
&\Rightarrow f'_i(a'_i) = \beta_i(b_i) \\
&\Rightarrow 0 = g'_i f'_i(a'_i) = g'_i \beta_i(b_i) = \gamma_i g_i(b_i) \\
&\Rightarrow 0 = g_i(b_i) \\
&\Rightarrow b_i \in \operatorname{Ker} g_i = \operatorname{Im} f_i \\
(\star) &\Rightarrow b_i = f_i(a_i) \\
&\Rightarrow f'_i(a'_i) = \beta_i(b_i) = \beta_i f_i(a_i) = f'_i \alpha_i(a_i) \\
&\Rightarrow a'_i - \alpha_i(a_i) \in \operatorname{Ker} f'_i = \operatorname{Im} h'_{i-1} \\
&\Rightarrow a'_i - \alpha_i(a_i) = h'_{i-1}(c'_{i-1}) = h'_{i-1}\gamma_{i-1}(c_{i-1}) = \alpha_i h_{i-1}(c_{i-1}) \\
&\Rightarrow a'_i = \alpha_i[a_i + h_{i-1}(c_{i-1})].
\end{aligned}
$$

Using $(\star)$ and the fact that $f_i h_{i-1} = 0$, we also see that

$$b_i = f_i[a_i + h_{i-1}(c_{i-1})].$$

Hence $(a'_i, b_i) \in \operatorname{Im} \varphi_i$.

Finally, for the exactness at A_i, if $a_i \in \operatorname{Ker} \varphi_i$ then $a_i \in \operatorname{Ker} \alpha_i$ and $a_i \in \operatorname{Ker} f_i$. If $a_i \in \operatorname{Ker} f_i = \operatorname{Im} h_{i-1}$ then we have

$$a_i = h_{i-1}(c_{i-1}) = h_{i-1}\gamma_{i-1}^{-1}(c'_{i-1}).$$

Now if $a_i \in \operatorname{Ker} \alpha_i$ then

$$0 = \alpha_i(a_i) = \alpha_i h_{i-1}\gamma_{i-1}^{-1}(c'_{i-1}) = h'_{i-1}\gamma_{i-1}\gamma_{i-1}^{-1}(c'_{i-1}) = h'_{i-1}(c'_{i-1})$$

which gives $c'_{i-1} \in \operatorname{Ker} h'_{i-1} = \operatorname{Im} g'_{i-1}$ so

$$c'_{i-1} = g'_{i-1}(b'_{i-1}).$$

Thus we have that

$$a_i \in \operatorname{Ker} \varphi_i \Longrightarrow a_i = h_{i-1}\gamma_{i-1}^{-1} g'_{i-1}(b'_{i-1}) \in \operatorname{Im} h_{i-1}\gamma_{i-1}^{-1} g'_{i-1}.$$

4.34 Let $R = \mathbb{Z} \times \mathbb{Z}$ and consider the ring R as an R–module. It is free, with basis $\{(1,1)\}$. Let $M = \mathbb{Z} \times \{0\}$. Then M is a submodule of R. But M is not free, since from

$$(0,1)(x,0) = (0,0)$$

we see that no non-empty subset of M can be linearly independent.

4.35 If f is a monomorphism then since f is left cancellable we cannot have $f \circ g = 0$ with $g \neq 0$. Hence f is not a left zero divisor in $\text{End}_R(M)$.

Suppose now that M is free, with basis $\{m_i \mid i \in I\}$. We obtain the converse by showing that $\text{Ker}\, f \neq \{0\}$ implies that f is not a left zero divisor. Suppose then that $\text{Ker}\, f \neq \{0\}$. Let $(n_i)_{i\in I}$ be a family of non-zero elements of $\text{Ker}\, f$ and let $g : M \to M$ be the (unique) R-morphism such that $g(m_i) = n_i$ for every $i \in I$. Since $\{m_i \mid i \in I\}$ is a basis of M it is clear that

$$\{0\} \neq \text{Im}\, g \subseteq \text{Ker}\, f$$

whence $f \circ g = 0$ and f is not a left zero divisor.

4.36 If f is an epimorphism then since f is right cancellable we cannot have $g \circ f = 0$ with $g \neq 0$. Hence f is not a right zero divisor in $\text{End}_R(M)$.

The $\mathbb{Z}$-module $\mathbb{Z}$ is free. Multiplication by 2 is in $\text{End}_{\mathbb{Z}}(\mathbb{Z})$ and is clearly neither a right zero divisor nor an epimorphism.

4.37 Suppose that each P_i in the family $(P_i)_{i\in I}$ is projective. To prove that $\bigoplus_{i\in I} P_i$ is projective, consider the diagram

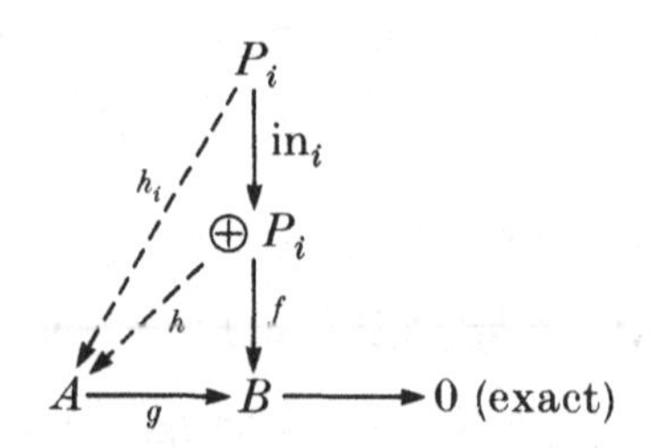

Since P_i is projective there exists an R-morphism $h_i : P_i \to A$ such that $g \circ h_i = f \circ \text{in}_i$. By the definition of coproduct, there is a unique R-morphism $h : \bigoplus_{i\in I} P_i \to A$ such that $h \circ \text{in}_i = h_i$. We show that $g \circ h = f$ whence $\bigoplus_{i\in I} P_i$ is projective.

Now $g \circ h \circ \text{in}_i = g \circ h_i = f \circ \text{in}_i$ and so, since every $x \in \bigoplus_{i\in I} P_i$ can be written uniquely in the form

$$x = \sum_{j\in J} \text{in}_j(p_j)$$

where J is a finite subset of I and $p_j \in P_j$ for all $j \in J$, we have

$$\begin{aligned}(g \circ h)(x) &= (g \circ h)\Big(\sum_{j \in J} \mathrm{in}_j(p_j)\Big) \\ &= \sum_{j \in J}(g \circ h \circ \mathrm{in}_j)(p_j) \\ &= \sum_{j \in J}(f \circ \mathrm{in}_j)(p_j) \\ &= f\Big(\sum_{j \in J} \mathrm{in}_j(p_j)\Big) \\ &= f(x)\end{aligned}$$

whence $g \circ h = f$.

4.38 If P_1 and P_2 are projective then (as seen in the previous question) so is $P_1 \oplus P_2$. We first observe that $P(M,N) \neq \emptyset$ since it clearly contains the zero morphism from M to N. Now given $f, g \in P(M,N)$ we have to show that $f - g \in P(M,N)$. For this purpose, consider the following diagram in which f factors through the projective module P_1 and g factors through the projective module P_2. Here π_1, π_2 are the projections of $P_1 \oplus P_2$ onto P_1, P_2 respectively.

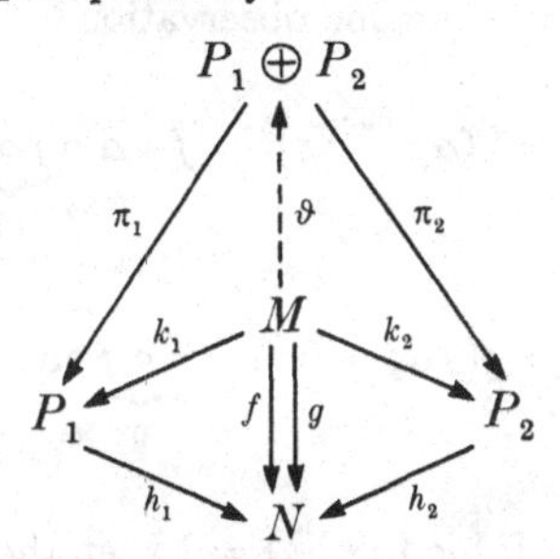

It follows from the definition of coproduct that there is a unique R-morphism $\vartheta : M \to P_1 \oplus P_2$ such that $\pi_1 \circ \vartheta = k_1$ and $\pi_2 \circ \vartheta = k_2$. Since then

$$h_1 \circ \pi_1 \circ \vartheta = h_1 \circ k_1 = f$$
$$h_2 \circ \pi_2 \circ \vartheta = h_2 \circ k_2 = g$$

we see that

$$f - g = (h_1\pi_1 - h_2\pi_2) \circ \vartheta$$

i.e. $f - g$ factors through the projective module $P_1 \oplus P_2$. Thus we have that $f - g \in P(M,N)$.

4.39 Since P', P'' are projective so also is $P' \oplus P''$. The sequence

$$0 \longrightarrow P' \xrightarrow{i} P' \oplus P'' \xrightarrow{\pi} P'' \longrightarrow 0$$

where $i(p') = (p', 0)$ and $\pi(p', p'') = p''$ is then split exact. If we let $j : P' \oplus P'' \to P'$ be a left hand splitting morphism and $\overline{\beta}$ a projective lifting of β then the given diagram can be extended to the following :

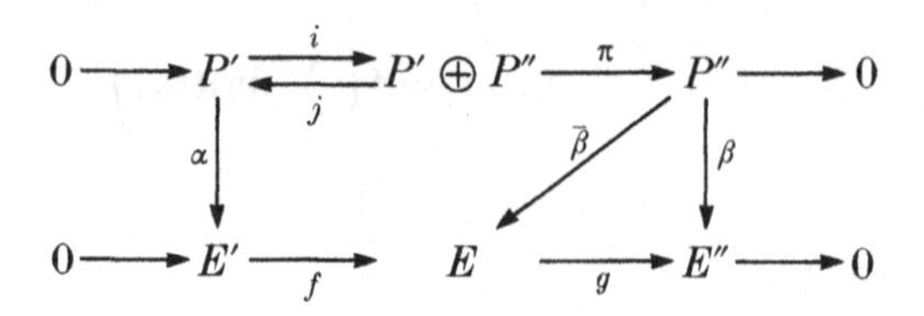

Consider the mapping $\gamma : P' \oplus P'' \to E$ given by

$$\gamma = f\alpha j + \overline{\beta}\pi.$$

It is clear that γ is an R–morphism. That it makes the diagram commutative follows from the observations

$$\gamma i = (f\alpha j + \overline{\beta}\pi)i = f \circ \alpha \circ \underbrace{j \circ i}_{\text{id}} + \overline{\beta} \circ \underbrace{\pi \circ i}_{0} = f\alpha;$$

$$g\gamma = g(f\alpha j + \overline{\beta}\pi) = \underbrace{g \circ f}_{0} \circ \alpha \circ j + \underbrace{g \circ \overline{\beta}}_{\beta} \circ \pi = \beta\pi.$$

4.40 (1) $\Longleftrightarrow$ (2) : If h.c.f.$\{r, \frac{n}{r}\} = 1$ then there exist integers x, y such that $xr + y\frac{n}{r} = 1$ whence, on passing to quotients,

$$r(x + n\mathbb{Z}) + \frac{n}{r}(y + n\mathbb{Z}) = 1 + n\mathbb{Z}$$

and therefore, for all $t \in \mathbb{Z}$,

$$t + n\mathbb{Z} = r(xt + n\mathbb{Z}) + \frac{n}{r}(yt + n\mathbb{Z}).$$

Also, we have that

$$\begin{aligned}
\frac{n}{r}(x+n\mathbb{Z}) = r(y+n\mathbb{Z}) &\iff \frac{n}{r}x - ry \in n\mathbb{Z} \\
&\iff \frac{n}{r}x - ry = \alpha n \\
&\iff \frac{n}{r}|ry \quad (\text{since } \frac{n}{r}|n) \\
&\iff \frac{n}{r}|y \quad (\text{since h.c.f.}\{\frac{n}{r}, r\} = 1) \\
&\iff ry \in n\mathbb{Z} \\
&\iff r(y+n\mathbb{Z}) = 0 + n\mathbb{Z}.
\end{aligned}$$

Thus if (2) holds we have that

$$\mathbb{Z}/n\mathbb{Z} \simeq \frac{n}{r}(\mathbb{Z}/n\mathbb{Z}) \oplus r(\mathbb{Z}/n\mathbb{Z})$$

and the sequence splits.

Conversely, if the sequence splits then we have the above isomorphism and so there exist $x, y \in \mathbb{Z}$ such that

$$r(x+n\mathbb{Z}) + \frac{n}{r}(y+n\mathbb{Z}) = 1 + n\mathbb{Z}.$$

It follows that $rx + \frac{n}{r}y = 1 + \alpha n$ for some $\alpha \in \mathbb{Z}$. This can be written

$$rx + (y - \alpha r)\frac{n}{r} = 1$$

and shows that h.c.f.$\{r, \frac{n}{r}\} = 1$.

(2) $\iff$ (3) : If (2) holds then $r(\mathbb{Z}/n\mathbb{Z})$ is a direct summand of the free $\mathbb{Z}/n\mathbb{Z}$-module $\mathbb{Z}/n\mathbb{Z}$ (a basis of which is $\{1+n\mathbb{Z}\}$); hence (3) holds. Conversely, if (3) holds then, by the definition of projective module, the sequence splits.

Taking $n = 6$ and $r = 3$ we have that (2) is satisfied and so it follows that $\mathbb{Z}/2\mathbb{Z} \simeq 3(\mathbb{Z}/6\mathbb{Z})$ is a projective $\mathbb{Z}/6\mathbb{Z}$-module. However, it cannot be free; for a free $\mathbb{Z}/6\mathbb{Z}$-module is a direct sum of copies of $\mathbb{Z}/6\mathbb{Z}$ and so has at least six elements, whereas $\mathbb{Z}/2\mathbb{Z}$ has only two.

4.41 That $\Delta_{A,B} = \{f \in \mathrm{Mor}_R(X,Y) \mid f^{\rightarrow}(A) \subseteq B\}$ is a submodule of the R-module $\mathrm{Mor}_R(X,Y)$ follows immediately from the observations

(1) $\Delta_{A,B} \neq \emptyset$ since it contains the zero morphism;
(2) $f, g \in \Delta_{A,B} \Rightarrow (f+g)^{\rightarrow}(A) \subseteq B \Rightarrow f + g \in \Delta_{A,B}$;
(3) $f \in \Delta_{A,B} \Rightarrow \lambda f \in \Delta_{A,B}$.

Suppose now that $f \in \Delta_{A,B}$. Then since $f^{\rightarrow}(A) \subseteq B$ there is a unique R–morphism $f_\star : X/A \to Y/B$ such that the diagram

$$\begin{array}{ccc} X & \xrightarrow{f} & Y \\ \downarrow{\natural_A} & & \downarrow{\natural_B} \\ X/A & \xrightarrow[f_\star]{} & Y/B \end{array}$$

is commutative. We can therefore define a mapping

$$\varsigma : \Delta_{A,B} \to \mathrm{Mor}_R(X/A, Y/B)$$

by the prescription $\varsigma(f) = f_\star$. It is readily verified that ς is an R–morphism; in fact, by the uniqueness of $f_\star$ we have $(f+g)_\star = f_\star + g_\star$ and $(\lambda f)_\star = \lambda f_\star$.

To see that ς is surjective, let $t \in \mathrm{Mor}_R(X/A, Y/B)$ and consider the diagram

$$\begin{array}{ccc} X & & Y \\ \downarrow{\natural_A} & \searrow^{t\circ\natural_A} & \downarrow{\natural_B} \\ X/A & \xrightarrow[t]{} & Y/B \end{array}$$

Since X is projective there exists an R–morphism $g : X \to Y$ such that $\natural_B \circ g = t \circ \natural_A$; in fact, g is a projective lifting of $t \circ \natural_A$. It follows that $g^{\rightarrow}(A) \subseteq B$ and so $\varsigma(g) = t$.

Now examine $\mathrm{Ker}\,\varsigma$. We have

$$\begin{aligned} f \in \mathrm{Ker}\,\varsigma &\iff f_\star = 0 \\ &\iff (\forall x \in X)\ f(x) + B = 0 + B \\ &\iff (\forall x \in X)\ f(x) \in B \\ &\iff f^{\rightarrow}(X) \subseteq B \\ &\iff f \in \Delta_{X,B}. \end{aligned}$$

Thus, by the first isomorphism theorem,

$$\mathrm{Mor}_R(X/A, Y/B) = \mathrm{Im}\,\varsigma \simeq \Delta_{A,B}/\,\mathrm{Ker}\,\varsigma = \Delta_{A,B}/\Delta_{X,B}.$$

4.42 The diagram

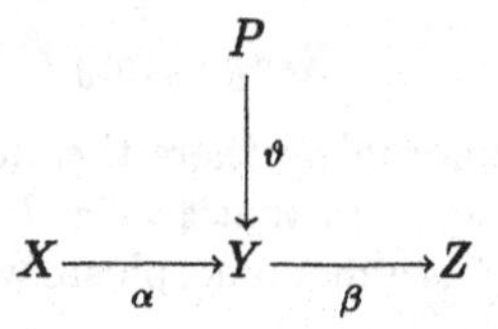

is given to be such that the row is exact and $\beta \circ \vartheta = 0$. It follows that

$$\operatorname{Im}\vartheta \subseteq \operatorname{Ker}\beta = \operatorname{Im}\alpha.$$

Let $\alpha^+ : X \to \operatorname{Im}\alpha$ and $\vartheta^+ : P \to \operatorname{Im}\alpha$ be the R–morphisms induced by α and ϑ. Then α^+ is an epimorphism and we have the diagram

$$\begin{array}{ccccc} & & P & & \\ & & \big\downarrow{\scriptstyle \vartheta^+} & & \\ X & \xrightarrow[\alpha^+]{} & \operatorname{Im}\alpha & \longrightarrow & 0 \end{array}$$

in which the row is exact. Since P is projective, there exists $\varsigma : P \to X$ such that $\alpha^+ \circ \varsigma = \vartheta^+$. Now let $i : \operatorname{Im}\alpha \to Y$ be the canonical inclusion. Then we have

$$\alpha \circ \varsigma = i \circ \alpha^+ \circ \varsigma = i \circ \vartheta^+ = \vartheta.$$

To apply induction in the second part of the question, note first that the existence of $k_1 : P_1 \to Q_1$ such that $h_1 \circ k_1 = k_0 \circ g_1$ follows from the fact that h_1 is an epimorphism and P_1 is projective.

Suppose, by way of induction, that for $t = 1, \ldots, n-1$ we have constructed $k_t : P_t \to Q_t$ such that $h_t \circ k_t = k_{t-1} \circ g_t$. Consider the diagram

$$\begin{array}{ccccc} P_n & \xrightarrow{g_n} & P_{n-1} & \xrightarrow{g_{n-1}} & P_{n-2} \\ & & \big\downarrow{\scriptstyle k_{n-1}} & & \big\downarrow{\scriptstyle k_{n-2}} \\ Q_n & \xrightarrow[h_n]{} & Q_{n-1} & \xrightarrow[h_{n-1}]{} & Q_{n-2} \end{array}$$

in which each of the rows is exact. We have

$$\begin{aligned} h_{n-1} \circ (k_{n-1} \circ g_n) &= (h_{n-1} \circ k_{n-1}) \circ g_n \\ &= k_{n-2} \circ g_{n-1} \circ g_n \\ &= k_{n-2} \circ 0 \\ &= 0 \end{aligned}$$

and so, P_n being projective, the first part of the question yields the existence of an R–morphism $k_n : P_n \to Q_n$ such that $h_n \circ k_n = k_{n-1} \circ g_n$. This then completes the inductive argument.

4.43 In the given diagram it is clear that $\natural \circ i \circ j = 0$ and so

$$\operatorname{Ker} g = \operatorname{Im} j \subseteq \operatorname{Ker} \natural \circ i.$$

Since g is an epimorphism there therefore exists a unique R–morphism $\vartheta : B \to P/\operatorname{Im} i \circ j$ such that $\natural \circ i = \vartheta \circ g$.

To prove that ϑ is a monomorphism we use the fact that g is surjective. Now

$$\begin{aligned}\vartheta[g(a)] = \vartheta[g(b)] &\Rightarrow (\natural \circ i)(a) = (\natural \circ i)(b)\\ &\Rightarrow i(a) - i(b) \in \operatorname{Im} i \circ j\\ &\Rightarrow i(a-b) = i[j(x)] \quad \text{for some } x \in \operatorname{Ker} g\\ &\Rightarrow a - b = j(x) \quad \text{since } i \text{ is monic}\\ &\Rightarrow g(a) - g(b) = g(a-b) = g[j(x)] = 0\end{aligned}$$

and the result follows.

If P is quasi-projective then the last part follows from the above on taking $A = P, B = P/N$ and $g = \natural_N$.

For the converse, consider the diagram

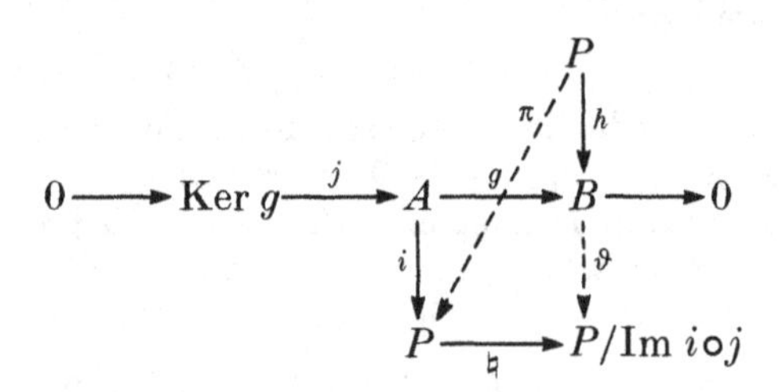

By the first part of the question there exists a unique monomorphism ϑ such that $\vartheta \circ g = \natural \circ i$. By hypothesis there exists a unique R–morphism π such that $\natural \circ \pi = \vartheta \circ h$. Let $p \in \operatorname{Im} \pi$. Then $p = \pi(q)$ so we can chase to get

$$\natural(p) = \natural[\pi(q)] = \vartheta[h(q)] = \vartheta[g(a)] = \natural[i(a)]$$

giving

$$p - i(a) \in \operatorname{Ker} \natural = \operatorname{Im} i \circ j$$

whence $p \in \operatorname{Im} i$. Thus we have that $\operatorname{Im} \pi \subseteq \operatorname{Im} i$. Since i is monic there therefore exists a unique R–morphism $\varsigma : P \to A$ such that $i \circ \varsigma = \pi$. Then

$$\vartheta \circ g \circ \varsigma = \natural \circ i \circ \varsigma = \natural \circ \pi = \vartheta \circ h$$

and, since ϑ is monic (by the first part of the question), $g \circ \varsigma = h$.

Test paper 1

Time allowed : 3 hours
(Allocate 20 marks for each question)

1 Let R be a ring and let N be the smallest subring of R that contains every nil two-sided ideal of R. Show that N is a nil two-sided ideal of R.

Suppose that I is a right ideal of R. Prove that $I + RI$ is a two-sided ideal of R. Prove also that, for any integer $m \geq 1$,

$$(I + RI)^m \subseteq I^m + RI^m.$$

Deduce that if I is a nilpotent right ideal of R then $I \subseteq N$.

Does N contain every nilpotent left ideal of R?

2 Let R be an integral domain in which it is impossible to find an infinite sequence (a_n) with the property that a_{n+1} is a proper factor of a_n for every n. Prove that R is a unique factorisation domain if and only if all the irreducible elements of R are prime.

Express $9 + 19i$ as a product of irreducibles in $\mathbb{Z}[i]$.

Show that in the quadratic domain $\mathbb{Z}[\sqrt{14}]$ the equations

$$14 = \sqrt{14} \cdot \sqrt{14} = 2 \cdot 7$$

do not violate unique factorisation.

3 Let F be a field and let f be a non-linear irreducible polynomial in $F[X]$. Show that $E = F[X]/(f)$ is an extension of F and that f has a linear factor $X - \alpha$ in $E[X]$ where $\alpha = X + (f) \in E$. Show also that $E = F(\alpha)$.

Now let $F = \mathrm{GF}(3) = \mathbb{Z}/(3)$ be the field with three elements and let $f(X) = X^3 - X + 1$.

(a) Show that f is irreducible over F.
(b) How many elements are there in $F(\alpha)$?
(c) Determine the inverse of α in $F(\alpha)$.
(d) Show that f splits completely into linear factors in $F(\alpha)[X]$ and find these factors.
(e) Describe the Galois group $\mathrm{Gal}(F(\alpha), F)$.

4 Let M and N be R–modules and $f : M \to N$ an R–morphism. If A, B are submodules of M, N respectively prove that the following statements are equivalent :

(1) $f^{\to}(A) \subseteq B$;

(2) there is a unique R–morphism $f_\star : M/A \to N/B$ such that the diagram

$$\begin{array}{ccc} M & \xrightarrow{f} & N \\ \downarrow{\natural_A} & & \downarrow{\natural_B} \\ M/A & \xrightarrow[f_\star]{} & N/B \end{array}$$

is commutative.

Show further that such an R–morphism $f_\star$, when it exists, is

(a) a monomorphism if and only if $A = f^{\leftarrow}(B)$;
(b) an epimorphism if and only if $N = B + \mathrm{Im}\, f$.

Given the diagram $M \xrightarrow{f} N \xrightarrow{g} P$ of R–modules and R–morphisms show that, with appropriate definitions of the morphisms involved, the sequence

$$0 \to \mathrm{Ker}\, f \to \mathrm{Ker}\, g \circ f \to \mathrm{Ker}\, g \to N/\,\mathrm{Im}\, f \to P/\,\mathrm{Im}\, g \circ f \to P/\,\mathrm{Im}\, g \to 0$$

is exact.

5 Let $A_1, \ldots, A_m$ and $B_1, \ldots, B_n$ be R–modules. For each R–morphism $f : \bigoplus_{k=1}^m A_k \to \bigoplus_{k=1}^n B_k$ define

$$f_{ji} = \mathrm{pr}_j^B \circ f \circ \mathrm{in}_i^A$$

where $\mathrm{pr}_j^B : \bigoplus_{k=1}^n B_k \to B_j$ is the canonical epimorphism and $\mathrm{in}_i^A : A_i \to \bigoplus_{k=1}^m A_k$ is the canonical monomorphism. If M denotes the set of $n \times m$ matrices $[\vartheta_{ji}]$ where each $\vartheta_{ji} : A_i \to B_j$ is an R–morphism, show that the mapping

$$\varsigma : \mathrm{Mor}_R\left(\bigoplus_{i=1}^{m} A_i, \bigoplus_{j=1}^{n} B_j\right) \longrightarrow M$$

described by $\varsigma(f) = [f_{ji}]$ is an abelian group isomorphism, so that f is uniquely determined by the $n \times m$ matrix $[f_{ji}]$.

Show also that the composite R–morphism

$$\bigoplus_{i=1}^{m} A_i \xrightarrow{\ f\ } \bigoplus_{j=1}^{n} B_j \xrightarrow{\ g\ } \bigoplus_{k=1}^{p} C_k$$

is represented by the matrix product $[g_{kj}][f_{ji}]$.

Hence establish, for every R-module A, a ring isomorphism

$$\mathrm{Mor}_R(A^m, A^m) \simeq \mathrm{Mat}_{m \times m}[\mathrm{Mor}_R(A, A)].$$

Test paper 2

Test paper 2

Time allowed : 3 hours
(Allocate 20 marks for each question)

1 Let R be a commutative ring with a 1 and let A be an ideal which is maximal in the set of ideals of R that are not finitely generated. Suppose, if possible, that $xy \in A$ with $x, y \notin A$. Show that (A, x) and $C = \{c \in R \mid cx \in A\}$ are finitely generated ideals of R. If

$$(A, x) = (a_1 + b_1 x, a_2 + b_2 x, \ldots, a_n + b_n x)$$

and $C = (c_1, c_2, \ldots, c_m)$ show that

$$A = (a_1, a_2, \ldots, a_n, c_1 x, c_2 x, \ldots, c_m x).$$

Deduce that A is a prime ideal.

Hence show, using Zorn's axiom, that if every prime ideal of a commutative ring with a 1 is finitely generated then every ideal is finitely generated.

2 Let $\omega = \frac{1}{2}(-1 + \sqrt{-3})$ be a primitive cube root of unity. Define

$$\mathbb{Z}[\omega] = \{a + b\omega \mid a, b \in \mathbb{Z}\}.$$

Show that $\mathbb{Z}[\omega]$ becomes a euclidean domain on defining a norm by

$$N(a + b\omega) = a^2 - ab + b^2.$$

[*Hint.* The proof is similar to the usual argument which shows that $\mathbb{Z}[\sqrt{n}]$ is a euclidean domain for $-2 \leq n \leq 3$.]

Prove that $\mathbb{Z}[\sqrt{-3}]$ is a subring of $\mathbb{Z}[\omega]$. Deduce that a subring of a unique factorisation domain need not be a unique factorisation domain.

3 Let K be a finite extension of the field F. Explain what is meant by saying that K is a normal extension of F.

Is $\mathbb{Q}(\sqrt{5})$ a normal extension of $\mathbb{Q}$? Is $\mathbb{Q}(\sqrt[3]{2})$ a normal extension of $\mathbb{Q}$?

Find the Galois group of the extension $\mathbb{Q}(\sqrt{2}, \sqrt{5})$ of $\mathbb{Q}$. Find the subgroups of this Galois group and determine the corresponding fixed fields.

4 If N is an R–module and $(M_i)_{i\in I}$ is a family of R–modules, establish the following isomorphisms of abelian groups (in which $\prod$ denotes cartesian products):

$$\text{(1)} \quad \mathrm{Mor}_R\left(\bigoplus_{i\in I} M_i, N\right) \simeq \prod_{i\in I} \mathrm{Mor}_R(M_i, N);$$

$$\text{(2)} \quad \mathrm{Mor}_R\left(N, \prod_{i\in I} M_i\right) \simeq \prod_{i\in I} \mathrm{Mor}_R(N, M_i).$$

Deduce that if $(M_i)_{i\in I}$ and $(N_j)_{j\in J}$ are families of R–modules then there is an abelian group isomorphism

$$\mathrm{Mor}_R\left(\bigoplus_{i\in I} M_i, \prod_{j\in J} N_j\right) \simeq \prod_{\substack{(i,j)\\ \in I\times J}} \mathrm{Mor}_R(M_i, N_j).$$

Establish the abelian group isomorphism

$$\mathrm{Mor}_{\mathbb{Z}}(\mathbb{Z}, \mathbb{Z}) \simeq \mathbb{Z}.$$

If, for every positive integer n, the cartesian product of n copies of $\mathbb{Z}$ is denoted by $\mathbb{Z}^n$, deduce that, for positive integers n and m,

$$\mathrm{Mor}_{\mathbb{Z}}(\mathbb{Z}^n, \mathbb{Z}^m) \simeq \mathbb{Z}^{nm}.$$

5 If P is a projective R–module and if the diagram

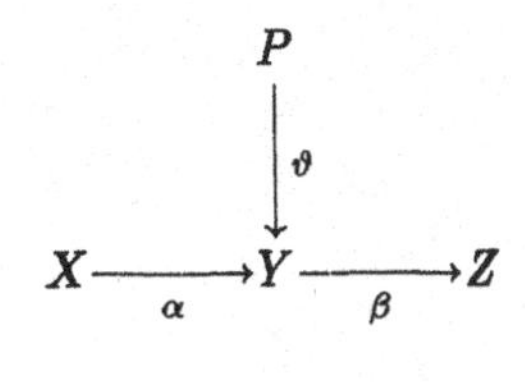

is such that the row is exact and $\beta \circ \vartheta = 0$, prove that there is an R-morphism $\varsigma : P \to X$ such that $\alpha \circ \varsigma = \vartheta$.

An exact sequence of the form

$$(\star) \qquad \cdots \longrightarrow P_n \xrightarrow{f_n} P_{n-1} \xrightarrow{f_{n-1}} \cdots \longrightarrow P_1 \xrightarrow{f_1} P_0 \longrightarrow 0$$

is said to *split* if there exist R-morphisms $g_i : P_i \to P_{i+1}$ such that

(1) $f_1 \circ g_0 = \mathrm{id}_{P_0}$;
(2) $(\forall i \geq 1)\ g_{i-1}f_i + f_{i+1}g_i = \mathrm{id}_{P_i}$.

Prove by induction that if each P_i is projective then the sequence $(\star)$ splits.

Test paper 3

Time allowed : 3 hours
(Allocate 20 marks for each question)

1 Explain what is meant by a euclidean domain and prove that every euclidean domain is a principal ideal domain.

Which of the following are euclidean domains? Give a proof, or a counter-example, to justify your assertions.

(a) $\mathbb{Z}[X]$;
(b) $\mathbb{Q}[X]$;
(c) $\mathbb{Z}[\sqrt{3}]$;
(d) $\mathbb{Z}[i]$.

By showing that $(2, 1+\sqrt{-3})$ is not principal in $\mathbb{Z}[\sqrt{-3}]$ prove that $\mathbb{Z}[\sqrt{-3}]$ is not euclidean.

2 Let $a_0 = 2$ and for $n = 0, 1, 2, \dots$ let a_{n+1} be the positive square root of a_n. Define $K_n = \mathbb{Q}(a_n)$.

(a) Show that $K_n \subseteq K_{n+1}$.
(b) Show that $K_{n+1} = K_n(a_{n+1})$ and that $(K_{n+1} : K_n) \leq 2$.
(c) Show by induction that $a_{n+1} \notin K_n$.
(d) Deduce the value of $(K_n : \mathbb{Q})$ for $n \geq 1$.

3 Find the irreducible factors in $\mathbb{Q}[X]$ of

$$f(X) = X^4 - 2X^2 - 3.$$

Show that $\mathbb{Q}(i, \sqrt{3})$ is a splitting field for f over $\mathbb{Q}$. Show also that $(\mathbb{Q}(i, \sqrt{3}) : \mathbb{Q}) = 4$ and write down a basis for $\mathbb{Q}(i, \sqrt{3})$ over $\mathbb{Q}$. Find the minimum polynomial of $i+\sqrt{3}$ over $\mathbb{Q}$. Deduce that $\mathbb{Q}(i,\sqrt{3}) = \mathbb{Q}(i+\sqrt{3})$.

Find the minimum polynomial of $i+\sqrt{3}$ over

(a) $\mathbb{Q}(i)$;

(b) $\mathbb{Q}(\sqrt{3})$;

(c) $\mathbb{Q}(i\sqrt{3})$.

4 The diagram of R–modules and R–morphisms

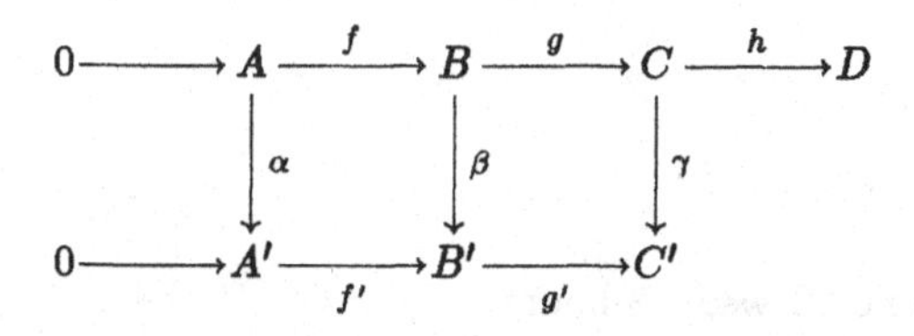

is given to be commutative with exact rows. If γ is an isomorphism, establish the exact sequence

$$0 \longrightarrow A \xrightarrow{\varphi} A' \oplus B \xrightarrow{\vartheta} B' \xrightarrow{\varsigma} D$$

where $\varsigma = h \circ \gamma^{-1} \circ g'$ and ϑ, φ are given by

$$\vartheta(a',b) = f'(a') - \beta(b), \qquad \varphi(a) = (\alpha(a), f(a)).$$

5 If $(M_\alpha)_{\alpha\in I}$ is a family of R–modules, explain what is meant by a ***product*** of this family. If $(P, (p_\alpha)_{\alpha\in I})$ and $(Q, (q_\alpha)_{\alpha\in I})$ are products of $(M_\alpha)_{\alpha\in I}$ prove that there is a unique R–morphism $\vartheta : P \to Q$ such that $q_\alpha \circ \vartheta = p_\alpha$ for every $\alpha \in I$.

Let $\left(\prod_{\alpha\in I} M_\alpha, (\mathrm{pr}_\alpha)_{\alpha\in I}\right)$ be the cartesian product of $(M_\alpha)_{\alpha\in I}$. A submodule M of $\prod_{\alpha\in I} M_\alpha$ is said to be a ***subdirect product*** of $(M_\alpha)_{\alpha\in I}$ if, for every $\alpha \in I$, the restriction $\mathrm{pr}_\alpha^M : M \to M_\alpha$ of the canonical projection pr_α is an R–epimorphism.

If N is an R–module and there is a family $(f_\alpha)_{\alpha\in I}$ of R–epimorphisms $f_\alpha : N \to M_\alpha$ such that

$$\bigcap_{\alpha\in I} \operatorname{Ker} f_\alpha = \{0\},$$

prove that N is isomorphic to a subdirect product of $(M_\alpha)_{\alpha\in I}$.

Deduce that $\mathbb{Z}$ is isomorphic to a subdirect product of $(\mathbb{Z}/n\mathbb{Z})_{n>1}$.

Test paper 4

Time allowed : 3 hours
(Allocate 20 marks for each question)

1 Let $A_1, \ldots, A_n$ be two-sided ideals of a ring R. We say that R is the *direct sum* of $A_1, \ldots, A_n$, and write $R = \bigoplus_{i=1}^n A_i$, if

(1) $R = \sum_{i=1}^n A_i$;
(2) $(i = 1, \ldots, n) \quad A_i \cap \sum_{j \neq i} A_j = \{0\}$.

If $S_1, \ldots, S_m$ are two-sided ideals of a ring R such that

(a) $R = \sum_{i=1}^m S_i$;
(b) no proper subset of $\{S_1, \ldots, S_m\}$ generates R;
(c) each S_i is a ring with a 1 containing no ideals other than $\{0\}$ and S_i,

show that $R = \bigoplus_{i=1}^m S_i$.

Now suppose that

$$R = \bigoplus_{i=1}^{m} S_i = \bigoplus_{i=1}^{n} T_i$$

are two representations of R as direct sums of two-sided ideals. For $i = 1, \ldots, m$ show that $S_i R = S_i$ and deduce that there is exactly one $j \in \{1, \ldots, n\}$ such that $S_i T_j = S_i$. Hence show that $S_i = T_j$ and deduce that the direct sum decompositions are essentially identical.

2 Let K be a subfield of $\mathbb{C}$ that is the splitting field of a polynomial $f \in \mathbb{Q}[X]$. By considering the Galois group of K over $\mathbb{Q}$, show that there are only finitely many distinct subfields of K containing $\mathbb{Q}$.

Let $D = \{(\mathbb{Q}(\xi) : \mathbb{Q}) \mid \xi \in K\}$ be the set of degrees of simple extensions of $\mathbb{Q}$ that are contained in K. Explain why D has a greatest

element, d say. If $\alpha \in K$ is such that $(\mathbb{Q}(\alpha) : \mathbb{Q}) = d$, show that $\mathbb{Q}(\alpha) = K$.
[*Hint.* If $\mathbb{Q}(\alpha) \neq K$ choose $\beta \in K \setminus \mathbb{Q}(\alpha)$ and consider the fields $\mathbb{Q}(\alpha + q\beta)$ where $q \in \mathbb{Q}$.]

3 Explain what is meant by saying that a finite extension of a field F is normal. Show that if $f \in F[X]$ and if K is the splitting field for f over F then K is a normal extension of F.

Let $F = \mathrm{GF}(p) = \mathbb{Z}_p$ and let $K = \mathrm{GF}(p^n)$, the splitting field for $X^{p^n} - X$ over F. Show that every irreducible polynomial of degree n in $F[X]$ splits completely over K.

Show that if $\alpha \in K$ then $\deg m_\alpha$ divides n, where m_α is the minimum polynomial of α over F. Deduce that no irreducible factor of $X^{p^n} - X$ in $F[X]$ has degree exceeding n.

Find the irreducible factors of $X^9 - X$ in $\mathrm{GF}(3)[X]$.

4 Prove that an R-module is noetherian (i.e. satisfies the ascending chain condition on submodules) if and only if every submodule of M is finitely generated.

Let M and N be noetherian R-modules and let P be an R-module such that there is a short exact sequence

$$0 \longrightarrow M \xrightarrow{\ f\ } P \xrightarrow{\ g\ } N \longrightarrow 0.$$

If A is a submodule of P show that $A \cap \operatorname{Im} f$ is finitely generated, say by $\{x_1, \ldots, x_r\}$. Now show that there exist $y_1, \ldots, y_n \in A$ such that

$$\{y_1, \ldots, y_n, x_1, \ldots, x_r\}$$

generates A. Deduce that P is also noetherian.

Hence show that if P is an R-module and M is a submodule of P then P is noetherian if and only if M and P/M are noetherian.

5 Prove that the following statements concerning an R-module P are equivalent :

(1) P is projective;
(2) every exact sequence $M \longrightarrow P \longrightarrow 0$ splits;
(3) P is a direct summand of a free R-module.

Let Δ_n be the ring of lower triangular $n \times n$ matrices $X = [x_{ij}]$ over a field F (so that $x_{ij} = 0$ if $i < j$). Let $A, B \in \Delta_n$ be given respectively by

$$a_{ij} = \begin{cases} 1 & \text{if } i = j+1; \\ 0 & \text{otherwise,} \end{cases} \qquad b_{ij} = \begin{cases} 1 & \text{if } i = j = 1; \\ 0 & \text{otherwise.} \end{cases}$$

If $\Theta_n = \{X \in \Delta_n \mid (i = 1, \ldots, n)\ x_{ii} = 0\}$ prove that

$$0 \longrightarrow \Delta_n B \xrightarrow{f} \Delta_n \xrightarrow{g} \Theta_n \longrightarrow 0,$$

where f is the canonical inclusion and g is given by $g(X) = XA$, is a split exact sequence of Δ_n–modules.

Deduce that Θ_n is Δ_n–projective.

Printed in the United States
By Bookmasters